LAKE NYASA CLIMATIC REGION FLORISTIC CHECKLIST

Roy E. Gereau, Asukile R. Kajuni,
Tim R. B. Davenport, and Henry J. Ndangalasi

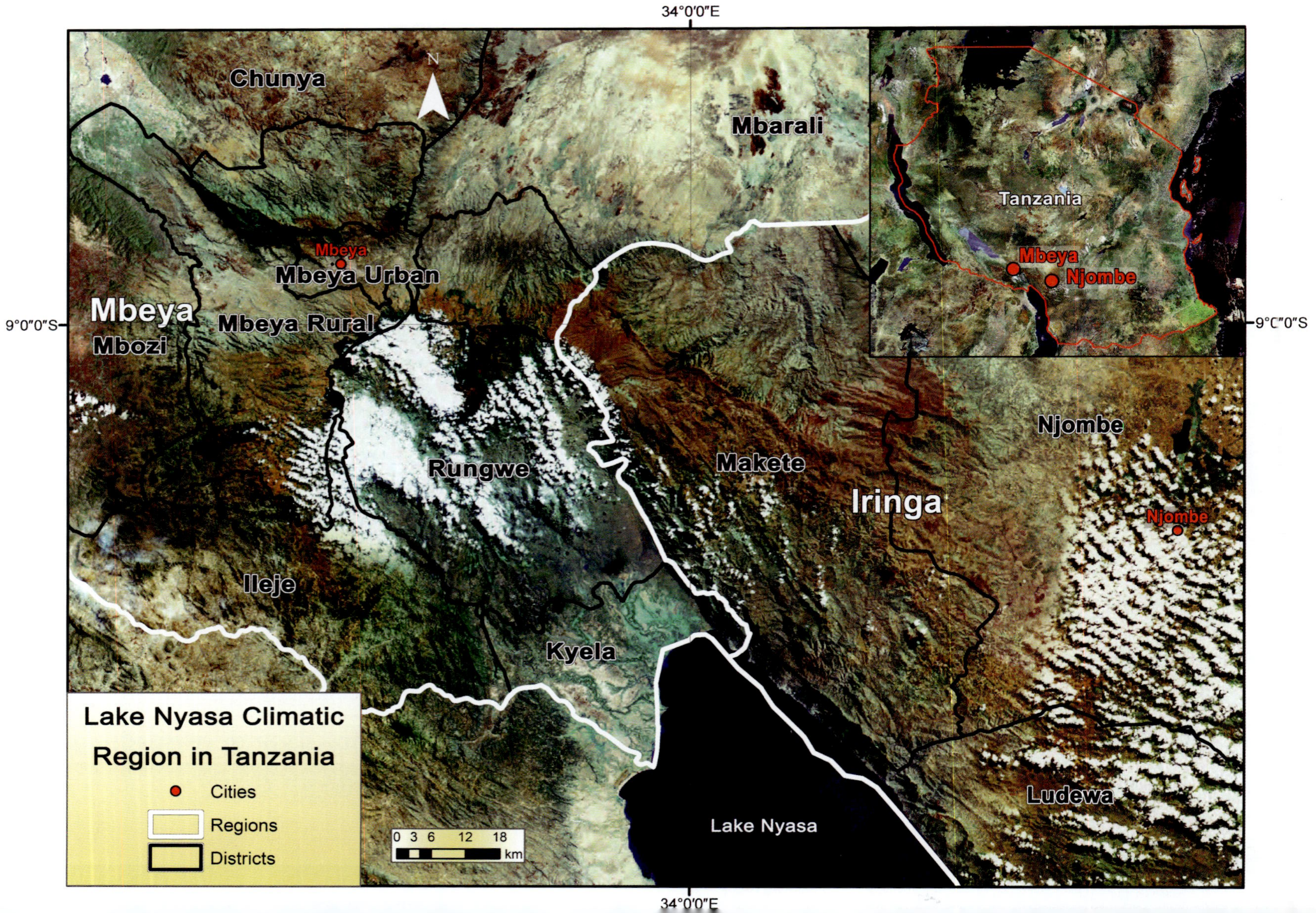
34°0'0"E
9°0"0"S
9°C"0"S
N
Chunya
Mbarali
Tanzania
Mbeya
Njombe
Mbeya
Mbeya Urban
Mbeya
Mbeya Rural
Mbozi
Rungwe
Makete
Iringa
Njombe
Njombe
Ileje
Kyela
Ludewa
Lake Nyasa
0 3 6 12 18
km
Lake Nyasa Climatic
Region in Tanzania
Cities
Regions
Districts

LAKE NYASA CLIMATIC REGION FLORISTIC CHECKLIST

Roy E. Gereau, Asukile R. Kajuni,
Tim R. B. Davenport, and Henry J. Ndangalasi

Missouri Botanical Garden

ISBN 978-1-935641-10-0
Library of Congress Control Number 2012930106

Monographs in Systematic Botany from the Missouri Botanical Garden, Volume 122
ISSN 0161-1542

Scientific Editor and Head: Victoria C. Hollowell
Associate Editor: Allison M. Brock
Associate Editor: Tammy M. Charron
Press Coordinator: Cirri R. Moran

P.O. Box 299
St. Louis, Missouri 63166-0299, U.S.A.
www.mbgpress.info

Photo credits—All cover photography by C. Davidson.
© http://www.floraoftheworld.com, 2008. Front cover (left): *Disa robusta* N. E. Br. (Orchidaceae), Kitulo Plateau, *W. Kindeketa 3024;* front cover (center): *Rothmannia engleriana* (K. Schum.) Keay (Rubiaceae), Chimala Scarp, *W. Kindeketa 2994;* front cover (right): *Impatiens cribbii* (Grey-Wilson) Grey-Wilson, Kitulo Plateau, *W. Kindeketa 3028;* back cover (upper left): *Savannosiphon euryphyllus* (Harms) Goldblatt & Marais (Iridaceae), Pungaluma Hills, *R. Gereau 7006;* back cover (upper right): *Globimetula anguliflora* (Engl.) Danser (Loranthaceae), Pungaluma Hills, *R. Gereau 6985;* back cover (lower left): *Bauhinia thonningii* Schumach. (Fabaceae–Caesalpinioideae), Pungaluma Hills, *R. Gereau 7056;* back cover (lower right): *Ensete ventricosum* (Welw.) Cheesman (Musaceae), Rungwe Mountain, *W. Kindeketa 2992.*

Frontispiece map: Parts of Iringa and Mbeya Regions, Tanzania, including study area (150-m resolution), with district boundaries, major cities, international boundary with Malawi, and north end of Lake Nyasa; inset map shows position of study area within Tanzania. Source: EarthSat NaturalVue Global Landsat Mosaic [DVD]. 2005. Environmental Systems Research Institute, Redlands, California.

ACKNOWLEDGMENTS

Roy E. Gereau is grateful to the National Geographic Society for Grant Number 4365-90, which supported the fieldwork that initiated this checklist, and to the Department of Botany, University of Dar es Salaam, for collaboration and logistical support. Tim R. B. Davenport and Henry J. Ndangalasi were funded by the Wildlife Conservation Society through the Southern Highlands Conservation Programme. We thank Charlotte Taylor for preparation of the maps, Burgund Bassüner and Tammy Charron for preparation of the frontispiece, Fred Keusenkothen for expert help with all aspects of image manipulation and presentation, and Jon Newell for heroic amounts of data entry. We particularly thank Henk Beentje, Alan Paton, and Iain Darbyshire for electronic copies of selected family treatments of the *Flora of Tropical East Africa.*

LAKE NYASA CLIMATIC REGION FLORISTIC CHECKLIST

Roy E. Gereau,[1] *Asukile R. Kajuni,*[2] *Tim R. B. Davenport,*[3] *and Henry J. Ndangalasi*[4]

Abstract

The Lake Nyasa Climatic Region described in this study is composed of the Livingstone, Kipengere, Mbeya, and Umalila ranges, Mporoto Ridge, Mt. Rungwe, and the Kitulo Plateau. The region is located on the rift escarpment at the northeast end of Lake Nyasa. Floristic studies in the area reveal many unique plants, some of which are endemic to this region. The climate of the region is governed by altitude and its proximity to Lake Nyasa, which results in generation of high convectional rainfall. The high rainfall and diversity of habitats interacting with soils and other environmental factors in the area are responsible for the richness of the flora, and current identifications of vascular plant taxa have yielded a total of 170 families, 733 genera, and 1736 species. The Lake Nyasa Climatic Region, therefore, has nearly 14% of the estimated total of 12,500 vascular plant species in all of Tropical East Africa (H. Beentje, pers. comm.). The largest families in the flora are Fabaceae (s. str. excluding Caesalpiniaceae and Mimosaceae; 156 species in 42 genera), Asteraceae (139/55), Poaceae (124/50), Rubiaceae (119/49), and Orchidaceae (97/28); the largest genera are *Crotalaria* L. (Fabaceae, 30 species), *Indigofera* L. (Fabaceae, 22 species), *Vernonia* Schreb. (Asteraceae, 21 species), *Cyperus* L. (Cyperaceae, 21 species), *Habenaria* Willd. (Orchidaceae, 18 species), and *Plectranthus* L'Hér. (Lamiaceae, 18 species). Taxa here reported as new to East Africa are *Cephalanthus natalensis* Oliv. (Rubiaceae), *Cerastium dregeanum* Fenzl (Caryophyllaceae), *Cynorkis symoensii* Geerinck & Tournay (Orchidaceae), *Dianthus zeyheri* Sond. subsp. *zeyheri* (Caryophyllaceae), *Panicum inaequilatum* Stapf & C. E. Hubb. (Poaceae), *Rhynchosia caribaea* (Jacq.) DC. var. *caribaea* (Fabaceae), *Tricalysia longituba* De Wild. var. *longituba* (Rubiaceae), *Tricalysia ngalaensis* Robbr. (Rubiaceae), and *Triumfetta angolensis* Sprague & Hutch. (Tiliaceae); new to Tanzania from elsewhere in East Africa are *Erica kingaensis* Engl. subsp. *bequaertii* (De Wild.) R. Ross (Ericaceae), *Peperomia fernandopoiana* C. DC. var. *butaguensis* (De Wild.) Düll (Piperaceae), and *Raphidiocystis jeffreyana* A. Fern. & R. Fern. (Cucurbitaceae).

The Lake Nyasa Climatic Region is located in the Iringa and Mbeya regions of Tanzania (frontispiece), and covers approximately 1200 km^2 (120,000 hectares) between latitude 8°50' and 9°32'S and longitude 33°15' and 34°09'E (Fig. 1). It extends to the southwest from the Eastern Arc Mountains of East Africa (Fig. 2), which are characterized by small and fragmented mountain blocs each having a patchwork of dense tropical forests and montane vegetation. It encompasses Kitulo National Park, Mpanga/Kipengere Game Reserve, and the following gazetted forest reserves: Mt. Rungwe, Chimala Scarp, Umalila, Sawago, Mporoto Ridge, Irenga, Irungu, Chuvwi, Masukulu, Kabulo, Kyosa, Mswima, Iyondo, Mpara, Mbeya Range, North Usafwa, Ileje Range, and Kyejo. The Lake Nyasa Climatic Region lies on the junction of the eastern and western arms of the Great Rift Valley of Africa, which runs from Mozambique through Malawi, Tanzania, Kenya, Ethiopia, and Eritrea into Saudi Arabia.

A closer scrutiny of the study area reveals many similarities to the Eastern Arc Mountains, which include the Udzungwa, Uluguru, Mahenge, Malundwe, Rubeho, Ukaguru, Nguru, Nguu, East Usambara, West Usambara, South Pare, and North Pare mountains in Tanzania, and the Taita Hills in Kenya (Platts et al., 2011). These discontinuous blocs of mountains described as "Galapagos Islands of Africa" have recently been a taxonomist's paradise, and new discoveries are being made regularly. Although our study area forms a southern extension of this ecologically important zone (Burgess et al., 2004), it has only recently started receiving attention. It is possible that the climatic influences associated with altitude and Lake Nyasa are responsible for the botanical differences between this area and the Eastern Arc Mountains (Fig. 2). However, recent and ongoing work by the Wildlife Conservation Society (Davenport, 2004, 2005, 2006; Jones et al., 2005; Davenport et al., 2006) has shown that this separation is not as marked

[1] Missouri Botanical Garden, P.O. Box 299, St. Louis, Missouri 63166-0299, U.S.A. roy.gereau@mobot.org.

[2] WWF Tanzania Programme Office, P.O. Box 63117, Dar es Salaam, Tanzania. akajuni@wwftz.org, ndemeleasu@yahoo.com.

[3] Wildlife Conservation Society, P.O. Box 922, Zanzibar, Tanzania. tdavenport@wcs.org.

[4] Botany Department, University of Dar es Salaam, P.O. Box 35060, Dar es Salaam, Tanzania. hjndangalasi@udsm.ac.tz; hjndangalasi@yahoo.com.

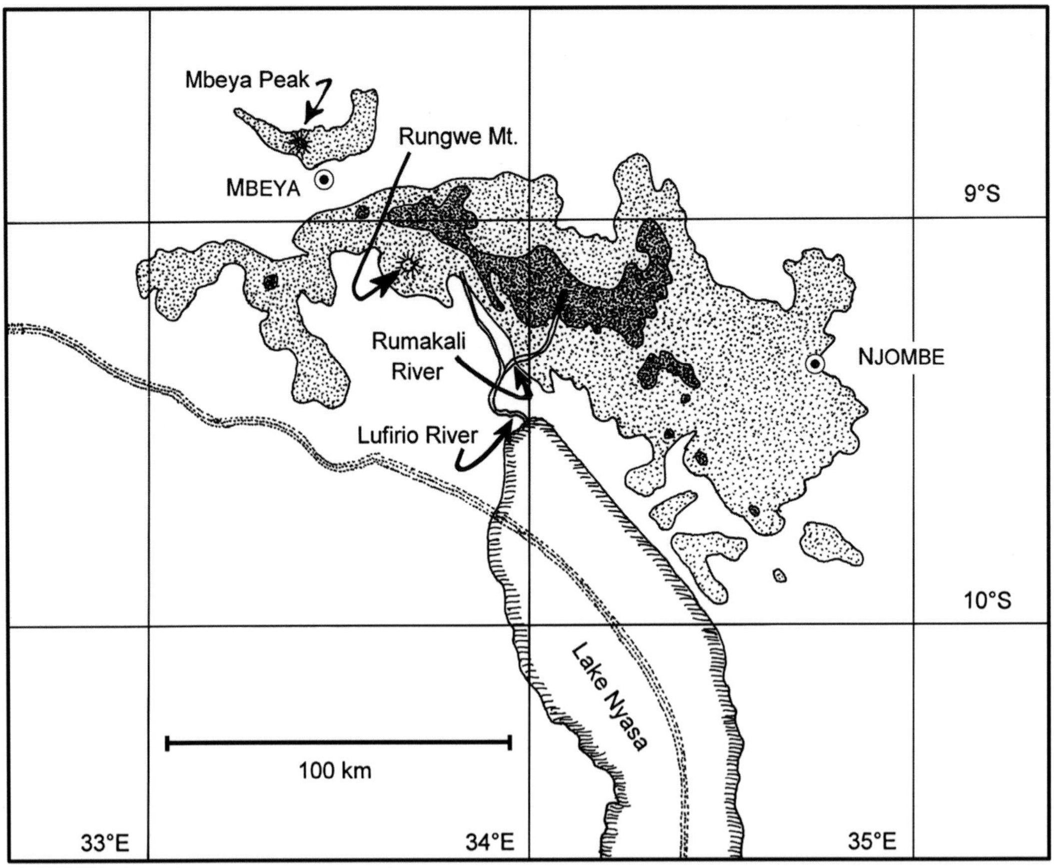

Figure 1. Map of the study area: The Lake Nyasa Climatic Region of southwestern Tanzania, with selected localities indicated. Lower contour line at 1500 m above sea level; upper contour line at 2000 m; lightly stippled area, land at 1500–2000 m; darkly stippled area, land above 2000 m; triple-dotted line, border between Tanzania (on north and east, above and to right) and Malawi (on south and west, below and to left).

zoologically as botanically. From the Makambako Gap the study area runs southwest to the northern tip of Lake Nyasa and forms a continuous range of high-altitude mountain ranges that divide the water catchment area into two drainage patterns: the northbound flowing into the Indian Ocean and the southbound flowing into Lake Nyasa. Several rivers and streams fed by numerous tributaries drain the flanks of the ridge, including Ruaha, Rumakali, Kiwira, Lufirio, Mbaka, Chimala, Kimani, Songwe, and Mbarali.

Elton and Cotteril (1879) wrote the first account of the floristic richness of the Kitulo Plateau, originally known as the Elton Plateau. Thereafter in the 20th century, several workers made attempts to describe some major resources including vascular plants from part of the areas surrounding the plateau. However, interest in the Kitulo Plateau was particularly intense due to the uniqueness of its resources and to the area's great potential for intensive agricultural development, sheep and dairy farming. The following studies have been particularly useful in providing insights into the natural resources of the Kitulo Plateau and surrounding areas: Gillman (1927), Greenway (1955), Hedberg (1957), Kerfoot (1964), Napper (1965), Prins and Prins-Lampert (1974), Cribb and Leedal (1982), TISCO (1983), Lovett et al. (1994), Lovett and Prins (1994), Davenport (2002, 2006), Davenport and Ndangalasi (2003), Davenport and Bytebier (2004), and Davenport et al. (2006, 2008).

The Lake Nyasa Climatic Region is of great ecological and phytogeographic importance for a number of reasons. The Indian Ocean drainage from this area is the source of the Ruaha, Mbarali, and Chimala rivers and numerous other streams that feed and provide constant recharge to the Usangu wetlands and floodplains that are impor-

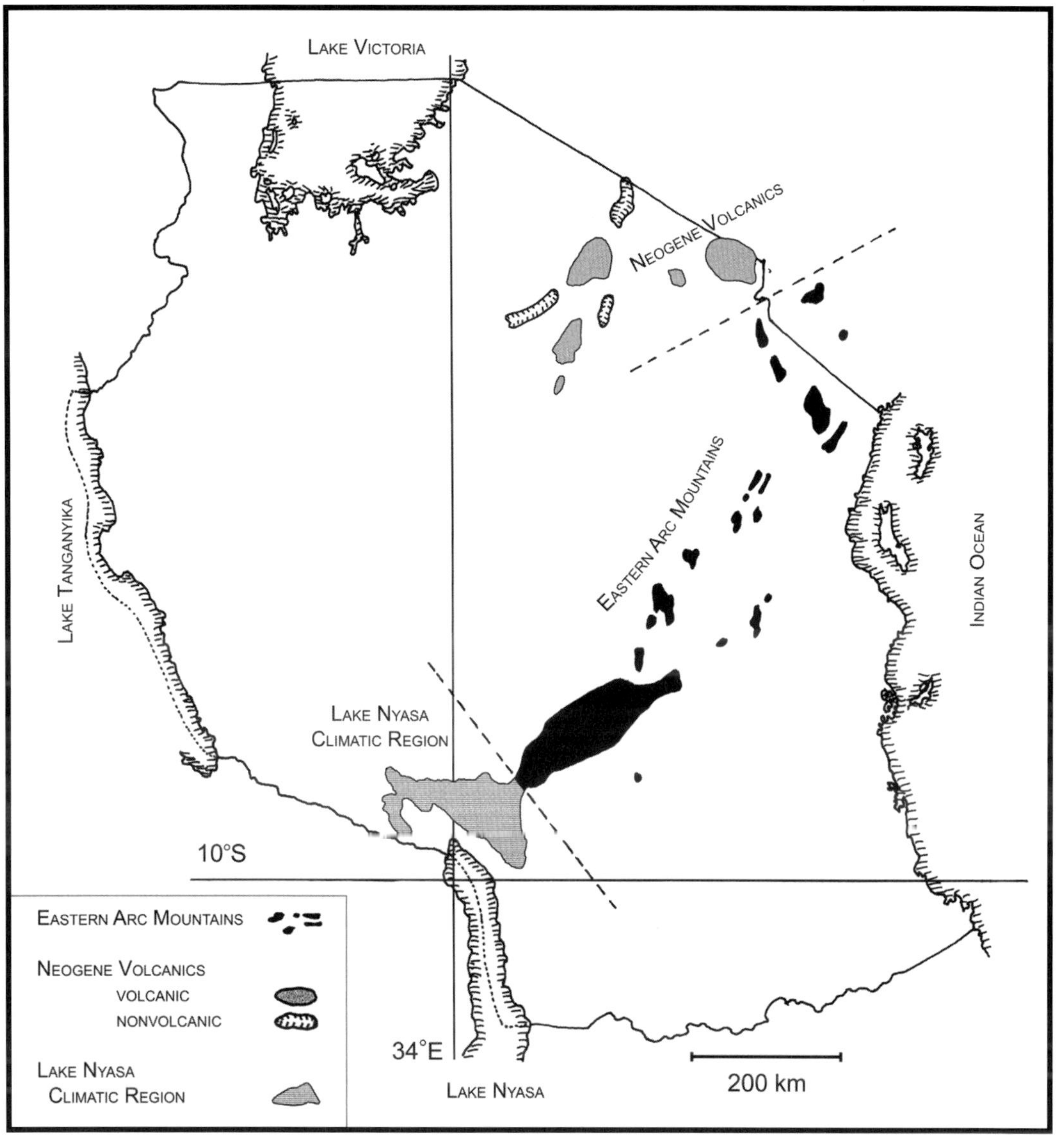

Figure 2. The Neogene Volcanics, Eastern Arc Mountains, and Lake Nyasa Climatic Region. Shaded areas between dashed lines indicate the Eastern Arc Mountains, with the Neogene Volcanics to the north and the Lake Nyasa Climatic Region to the southwest.

tant rice growing areas, producing over 60% of rice sold in urban market stalls in the country (Kamukala & Crafter, 1993). Much of the Usangu Plains have recently been gazetted as an extension of the Ruaha National Park and provide important refuge and grazing grounds for wildlife; the Usangu wetlands feed into the Great Ruaha River, whose waters are an important source of hydroelectric power, supplying over 60% of all electricity consumed in the country through the national grid from dams impounded at Mtera and Kidatu (Power et al., 1997; NORPLAN, 2002). The irrigation schemes for the Kilombero sugar plantations and rice farms are dependent on waters from the Ruaha River, and the whole Rufiji Delta with its rich marine and other aquatic resources depends on water resources from the Southern Highlands. The southern drainage from the region is the source of the Rumakali, Songwe, Kiwira, Mbaka, and Lufirio rivers, which water Rungwe District with its important cash

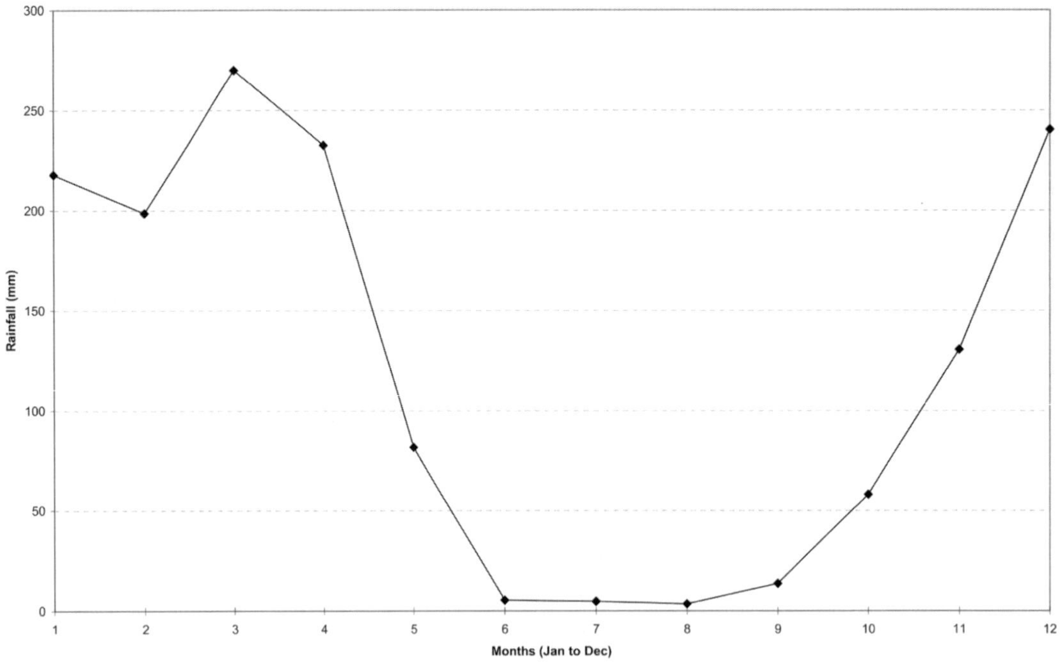

Figure 3. Average monthly precipitation for Kitulo Dairy Farm (1965–1994). Source: TISCO (1983), with additional data obtained from Kitulo Dairy Farm Meteorological Station in 1999.

crops of coffee, tea, avocados, and bananas; Kyela District with its high-quality and fragrant rice, abundant fruits, and cocoa; and Lake Nyasa with its high diversity of cichlid fish. The Kitulo Plateau has been described as harboring "one of the great floral spectacles of the world," where rolling montane grassland and patches of forest provide a unique environment and unparalleled botanical display with a plethora of globally significant flora and fauna (Davenport, 2002; Davenport & Bytebier, 2004). Management of natural and physical resources requires knowledge about the biology, ecology, and their interrelationship in space and time. Information about these resources is essential in order to guide decision-making processes about appropriate land uses and how best to manage and preserve them (Mwakilema & Davenport, 2005). Therefore, the Wildlife Conservation Society is working to understand better the flora, fauna, invertebrates, and other associated resources found in the region (Davenport, 2004, 2006), which is so critical at this point as a result of increased human populations and globalization of the economy. The open market economy has provided incentives for people to increase their agricultural produce through cultivating more land and encroaching on areas that originally were reserved for conservation purposes. Uncontrolled cultivation for agricultural crops results in loss of an important genetic pool from the original, indigenous flora due to change in the natural vegetation structure and composition. As more of the natural environment is converted to agricultural land, it means a loss in indigenous natural flora and fauna. The loss in indigenous natural resources through extinction is tantamount to a lost opportunity for science and human use, as we cannot recreate the natural world.

Climate of Study Area

Pratt and Gwynne (1977) described the climatic conditions of the study area as having one single season rainfall from November to May and a dry season from June to October. The mean annual rainfall is variable depending on altitude and proximity to Lake Nyasa but lies between 1200 and 2950 mm. The number of rainy days ranges between 117 in Kyela along Lake Nyasa and 159 days on the Kitulo Plateau (*Atlas of Tanzania*, 1967; TISCO, 1983). Long-term climatic data have been collected and kept at Kitulo Dairy Farm, which over a 30-year period show that December to April receives the heaviest rainfall, averaging 231.9 mm per month, and June to September receives the least amount, with average monthly precipitation below 10 mm (Fig. 3).

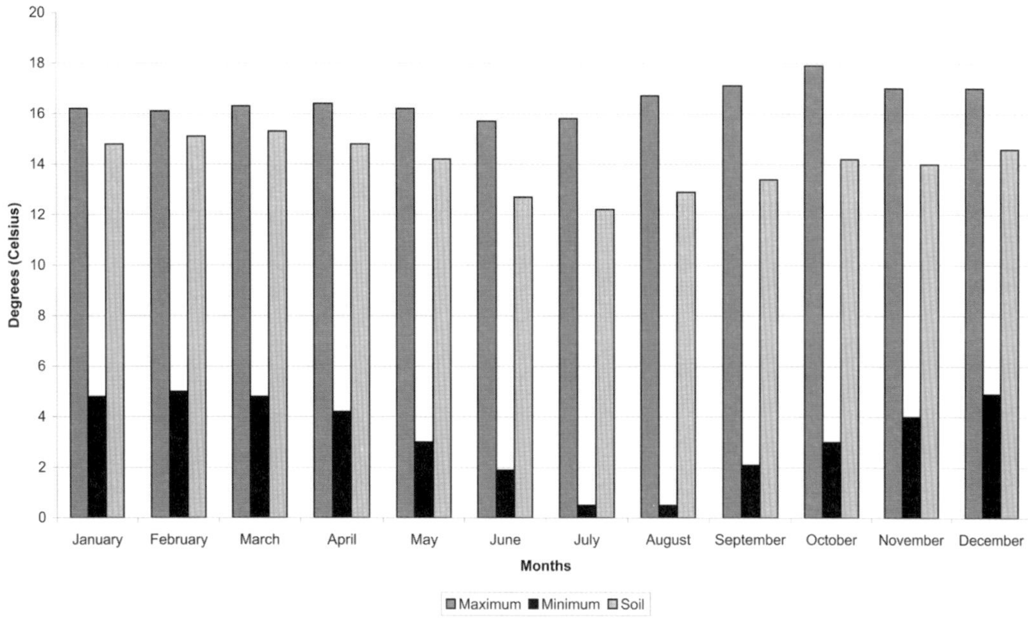

Figure 4. Mean monthly maximum and minimum and soil temperatures for Kitulo Dairy Farm (1965–1982). Source: TISCO (1983).

The plateau has a relatively uniform cool, moist climate, cool temperatures, low evapotranspiration rates, few sunshine hours per day, and brisk winds; the mean minimum and maximum temperatures are 3.1°C and 16.5°C, respectively. The soil temperatures show similar trends, with both air and soil temperatures being coolest in July and warmest in October (Fig. 4). During the cool season (June, July, and August), the air temperatures frequently dip below freezing and ground frosts occur almost every night. The lowest daily mean minimum temperature recorded at Kitulo was –9.5°C in July of 1973. Mean grass minimum temperatures are normally below 0°C for the months of June through September.

The maximum hours of sunshine and radiation occur during the dry, cool season, at which time the sky is relatively clear (Figs. 5 and 6). Solar radiation peaks in August, which is also the month of maximum daily sunshine hours (7.5 hours). Both hours of sunshine and radiation start to decrease in September and reach their minimum in January.

The weather records from Kitulo generally represent the climatic conditions obtained in the highlands and high-altitude areas within the Lake Nyasa Climatic Region. Kitulo Dairy Farm has maintained uninterrupted long-term meteorological data collection that is quite reliable and therefore used to illustrate the weather conditions found in the entire study area. However, localized differences resulting from physiographic conditions provide some exceptions to this general pattern. To the southwest of Mt. Rungwe, for example, convectional rain from Lake Nyasa continues throughout most of the year (Table 1).

Geology and Soils

Geologically, the rock formations in the study area are derived from the Ubendian system of stratigraphical units that have been aged in the Late Archaean era about 2000 to 1600 million years ago. Superimposed on the Ubendian Belt are zones of intense recent Proterozoic deformations (Early Proterozoic dates between 1600 million years and 750 million years), which have obscured and in some cases completely obliterated Late Archaean structures. These structures were further complicated by extensive continued faulting activity during the Karoo Rift episodes, where certain areas are still seismically active (Harpum, 1970). The Ubendian system is characterized by greater amounts of metamorphosed igneous rocks and by the presence of charnockitic rocks. For example, the rock formation of the Livingstone Mountains is of anorthositic composition with intrusive basic and ultra-basic complexes. As an example of the superimposition of more recent Proterozoic

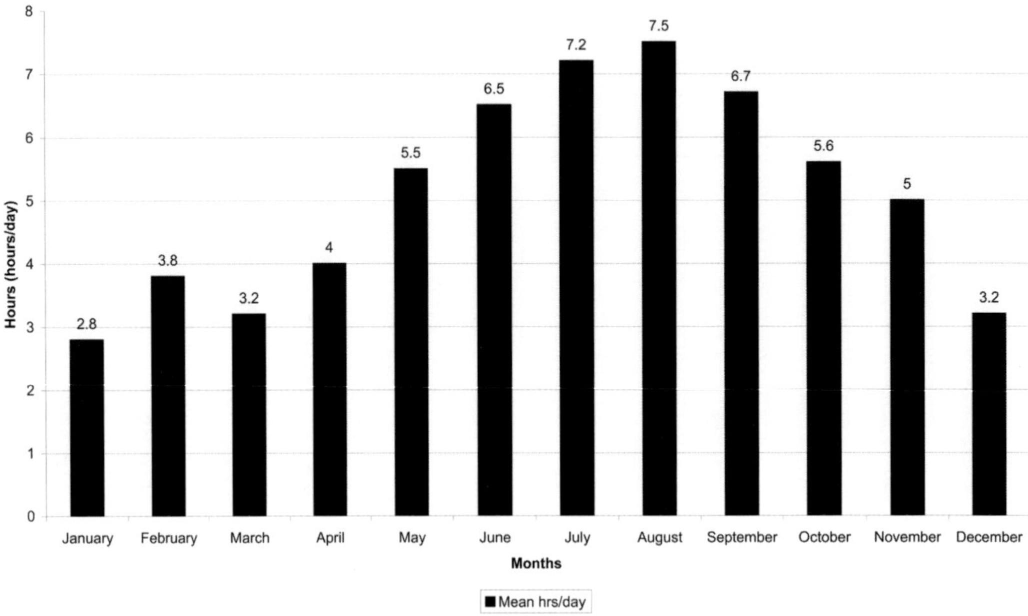

Figure 5. Mean daily sunshine (hours/day) for Kitulo Dairy Farm (1965–1982). Source: TISCO (1983).

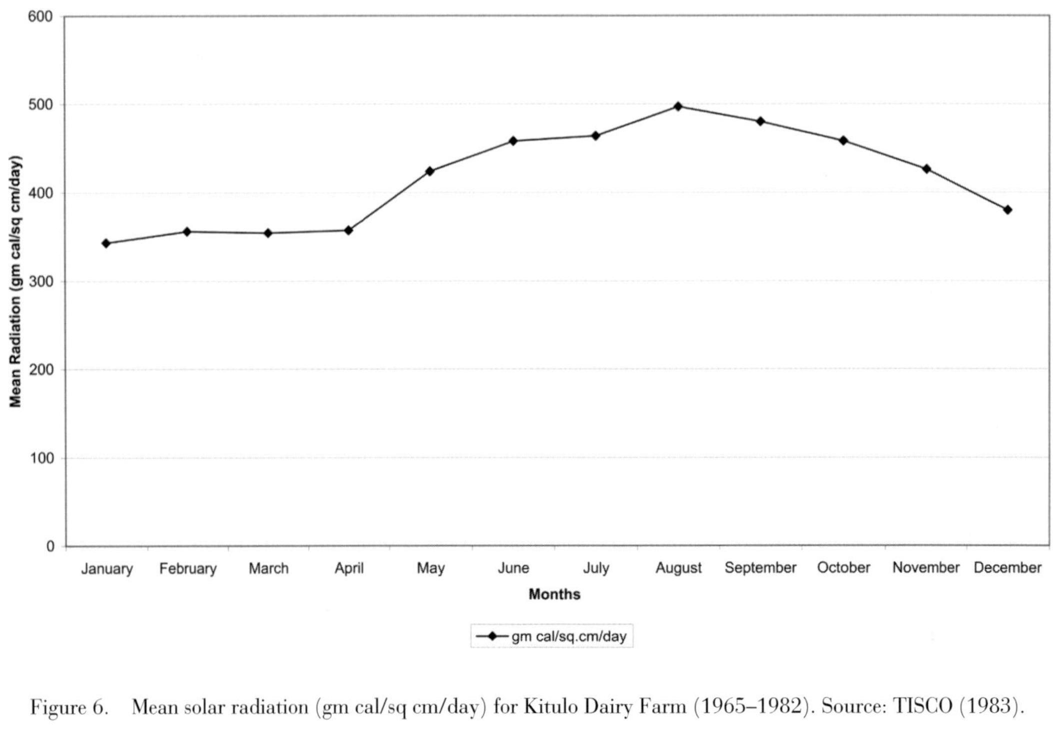

Figure 6. Mean solar radiation (gm cal/sq cm/day) for Kitulo Dairy Farm (1965–1982). Source: TISCO (1983).

Table 1. Maximum (Ma) and minimum (Mi) mean monthly temperature (°C) and mean monthly rainfall (R) in millimeters from southwest of Mt. Rungwe near Syukula.

	Ma	Mi	R
Jan	23	15	312
Feb	24	16	173
Mar	25	16	262
Apr	24	14	361
May	22	13	152
June	22	11	78
July	22	9	97
Aug	23	12	12
Sep	25	14	31
Oct	26	16	9
Nov	27	16	160
Dec	26	16	326

deformations, the soils of Kitulo Plateau and much of the adjacent areas in the region are derived from volcanic pumice as a result of volcanic eruptions from Rungwe and Ngosi volcanoes and are well drained. The plateau is composed of Pliocene basaltic and phonolitic lavas to an average depth of 150 m, which overlay pre-Karoo formations (Harkin, 1960). The soils range from humic ferrallitic on various parent materials to humic ferrisols of high-altitude regions (*Atlas of Tanzania*, 1967). In all, six principal soil types have been identified for Kitulo and are related to physiographic factors as shown in Table 2.

Ecology

Much of the area of the Lake Nyasa Climatic Region lies in eco-climatic zone II and a small area in zone I (Pratt & Gwynne, 1977). Zone I is characterized by an Afroalpine climate governed mostly by altitude and is dominated by Afroalpine moorland and montane grassland above the forest line, while eco-climatic zone II comprises highlands with a tropical climate. The Lake Nyasa Climatic Region has the largest and one of the most important Afromontane and Afroalpine grassland communities in the country (White, 1983). The land in this eco-climatic zone is the center of forest industry, although it also has large tracts under intensive subsistence agriculture or under cultivation for cash crops like tea and coffee. The dominant vegetation includes moist forests and derived grassland and bushland with or without glades. The indigenous montane and upper montane forests in these areas are the sources of economically important large forest trees including *Ficalhoa laurifolia* Hiern (Theaceae), *Juniperus procera* Hochst. ex Engl. (Cupressaceae), *Ocotea usambarensis* Engl. (Lauraceae), *Podocarpus latifolius* (Thunb.) R. Br. ex Mirb. (Podocarpaceae), and *Pouteria adolfi-friedericii* (Engl.) A. Meeuse subsp. *australis* (J. H. Hemsl.) L. Gaut. (Sapotaceae). There are also significant areas of bamboo forest dominated by *Sinarundinaria alpina* (K. Schum.) C. S. Chao & Renvoize (Poaceae). Bushland or evergreen shrubland are the commonest derivatives where shifting cultivation has been practiced or where forest has been disturbed but not burned with sufficient regularity to maintain grassland.

Collecting History

James Elton, the British consul to Mozambique, was probably the first botanical explorer of the Lake Nyasa Climatic Region (Moffett, 1958); he crossed the Kitulo Plateau in late October 1877 and published the first account of its floristic composition (Elton & Cotteril, 1879). German botanists, most notably W. Goetze and A. Stolz, carried out most subsequent plant collection in the area through the end of World War I; Engler and Goetze (1902) included Mt. Rungwe in their work on the montane vegetation of Tanzania. After World War I, the area received relatively sporadic attention from plant collectors (with the notable exception of H. J. Schlieben in Njombe District) until Mary Richards began her extensive collecting in the Mbeya and Njombe highlands in 1956, continuing through much of the 1960s (Cribb & Leedal, 1982). Many of the plant collectors working in the area from the 1970s until the present have been associated with the *Flora of Tropical East Africa* (see Beentje & Smith, 2001), the National Herbarium of Tanzania, the University of Dar es Salaam, the Missouri Botanical Garden, or the Wildlife Conservation Society. Despite a relatively large number of collections, floristic checklists and/or vegetation descriptions have been published only for Mbeya Peak (Kerfoot, 1964) and the Kitulo Plateau (Prins & Prins-Lampert, 1974; Lovett et al., 1994).

Conservation Status

Although much of the land area in the Lake Nyasa Climatic Region is gazetted as Forest Reserves and

Table 2. Kitulo Plateau Soil Types in relation to physiographic regions. Source: TISCO (1983).

Soil type	Physiographic region found
Black organic loam underlain by pumice	Undulating to hilly plateau
Black organic loam over very dark grey loam over pumice	Undulating to hilly plateau
Shallow, black organic loam over loamy pumice over clay loam	Rolling to steeply dissected highlands
Black peaty loam over pumice	Flat to almost flat floodplains
Shallow black loam over clay over weathered schistose bedrock	Hilly to steeply dissected highlands
Shallow, very dark brown stony loam over bedrock; very dark grey or black loam over bedrock	Steeply dissected to mountainous

theoretically accorded a legal protection status, there are serious problems that lead to continued loss of indigenous montane forests. Continued conversion to farmland and pine plantations and uncontrolled human encroachment and cultivation have completely altered the natural vegetation patterns. Some information is available on the extent of loss of montane forest habitat from illegal harvesting, fuelwood collection, and logging (Machaga et al., 2004, 2005), and this loss is also the subject of ongoing research; it is obvious that thinning and opening up of the forest has occurred in all areas. Some forests that were once almost impenetrable have been opened up in the last 30 years for potato and pyrethrum farming. The reduction in the size of these once intact indigenous montane forests poses serious threats to their capacity to ensure a stable, high-quality supply and recharge of water in the hydrological cycle. In addition, the indigenous montane forests play other important functions in the prevention of soil erosion, loss of topsoil, and damage from landslides, as well as providing valuable non-timber forest products and medicines for thousands of people (Machaga et al., 2004, 2005). Only the intact natural forest can sustain the intricate processes of water movements through vegetation and soil. The bryophyte and lichen epiphytes on woody plants are particularly important to ensure the interception and storage of large amounts of rainwater that are released slowly and recharge the water table. The Usangu Plain and Kilombero agricultural irrigation schemes, the Mtera and Kidatu hydroelectric power stations, and the Rufiji Basin depend entirely on the water from the Lake Nyasa Climatic Region, as do the Kyela rice schemes. Many long-term residents of the surrounding districts of Rungwe, Mbeya Rural, Mbeya Urban, Makete, Njombe, Ileje, and Kyela have witnessed changing hydrologic patterns and noted a reduction in the volume of water in most perennial rivers, and many once perennial streams have become intermittent.

Continued degradation of the area's high watershed values and the presence of many endemic and unique plants in the area have stimulated reconsideration of the protection status of many key areas. Following the gazetting of Kitulo National Park in 2004, it is now extremely important that other areas of this region be accorded better protection due to the current lack of management strategies and the increase in inappropriate land use practices.

Resource use associated with the increase in the human and livestock populations is often unsustainable, resulting in environmental degradation. There is a growing awareness of the importance of conservation, but progress takes time. The environmental management schemes in Makete and Njombe districts have yielded positive results, but are not practiced throughout the study area. However, in October 2002 the government, as a result of increased pressure arising from an increase of human activities resulting in unsustainable use of resources in the Usangu Plains and the surrounding areas, announced the gazettement of the Mpanga/Kipengere area as a protected area (Game Reserve). The Game Reserve covers an area of 1574.25 km^2 in part of the study area abutting the three districts of Njombe, Makete, and Mbarali (Staddon et al., 2003). The Mpanga/Kipengere Game Reserve comprises very important water-catchment areas.

The Kitulo Plateau, lying between the rugged peaks of the Kipengere, Uporoto, and Livingstone mountains, was gazetted as a national park in 2004 (Davenport, 2002; Davenport & Bytebier, 2004) in order to ensure the protection and preservation of an area that supports the largest and most important montane grassland community in Tanzania (Mwakilema & Davenport, 2005). Much of the park is located at ca. 2600 m on well-watered volcanic soils. However, the park also covers significant amounts of montane forest (Davenport et al., 2006). Kitulo National Park is located in the headwaters and forms the most important watersheds for the Great Ruaha River and is also well known for its floral significance.

Organization, Format, and Notation of Checklist

Families are listed alphabetically within the divisions Pteridophyta, Pinophyta (Gymnospermae), and Magnoliophyta (Angiospermae). Within Pteridophyta and Pinophyta, family circumscriptions follow Kubitzki (1990). Within Magnoliophyta, family circumscriptions for Liliopsida (Monocotyledonae) follow Dahlgren (1985); those for Magnoliopsida (Dicotyledonae) follow Cronquist (1981), with the exception of *Hebenstretia* L., which is here included in the Scrophulariaceae. Genera, species, and infraspecific taxa are listed alphabetically within families. One or more species appearing without specific epithets at the end of genera (e.g., "*Xysmalobium* sp." appearing twice under Asclepiadaceae) are known to be different from other species in the same genus cited in the checklist but have not been positively identified, and thus may be either new to science or disjunct from another geographic area and not yet reported from the Lake Nyasa Climatic Region.

A single reference is given to support the use of each species (or infraspecific taxon) name in the Lake Nyasa Climatic Region; this is not the original place of publication of the name unless that is the principal literature used for specimen identification or the only literature available. Publications other than major African floras are listed in full in the Literature Cited. Selected major African floras are cited using the following standard abbreviations: F.T.A. = Flora of Tropical Africa (Oliver, 1868–1877; Dyer, 1896–1913; Prain, 1916–1934; Hill, 1937; see also Stafleu & Cowan, 1981); F.T.E.A. = Flora of Tropical East Africa (Turrill & Milne-Redhead, 1952–1958; Hubbard & Milne-Redhead, 1958–1966; Milne-Redhead & Polhill, 1966–1972; Polhill, 1973–1978, 1982–1997; Beentje, 1998–2007, 2010; Beentje & Ghazanfar, 2005–2009; see also Beentje & Smith, 2001); F.W.T.A. = Flora of West Tropical Africa, second edition (Hutchinson & Dalziel, 1954, 1958; Hepper, 1963–1972); Fl. Zamb. = Flora Zambesiaca (Exell & Wild, 1960–1961; Exell et al., 1963, 1966; Brenan, 1970; Exell & Launert, 1970; Fernandes et al., 1971; Launert, 1978–1989; Launert & Pope, 1989–1991; Pope, 1992–1995, 1996–2005; Pope & Martins, 2005; Pope et al., 2006; Timberlake et al., 2007; Timberlake & Martins, 2008–2010).

Elevational ranges are taken directly from herbarium labels; a single number indicates that only one collection exists, or that only one label indicated elevation. Abbreviations cited for mountain ranges under "Distribution" in the checklist are as follows (Kipengere and Umalila ranges not included due to insufficient collections): KP = Kitulo Plateau; MP = Mbeya Range (including Mbeya Peak and Mt. Loleza); MR = Mporoto Ridge (including Ngozi Crater); NL = Northern Livingstone Mountains (i.e., north of Rumakali River); RM = Rungwe Mountain; SL = Southern Livingstone Mountains (i.e., south of Rumakali River).

A single representative specimen is cited for each species or infraspecific taxon unless the type specimen was collected within the Lake Nyasa Climatic Region, in which case the type and one other specimen (if any) are cited. Herbaria of deposit are indicated for type specimens. Most other cited specimens are deposited in DSM, EA, K, MO, or NHT.

Lake Nyasa Climatic Region Floristic Checklist

Pteridophyta

ASPLENIACEAE

1 genus and 9 species

Asplenium abyssinicum Fée; Beentje, F.T.E.A. (Aspleniaceae): 62. 2008.

Terrestrial herb. Montane forest. 1950–2400 m.

Distribution: MR—*Gereau 4620*

Asplenium aethiopicum (Burm. f.) Bech.; Beentje, F.T.E.A. (Aspleniaceae): 58. 2008.

Epiphytic or epilithic herb. Montane forest. 1920–2400 m.

Distribution: KP, MR, NL, RM, SL—*Gereau 3727*

Asplenium anisophyllum Kunze; Beentje, F.T.E.A. (Aspleniaceae): 36. 2008.

Epiphytic or epilithic herb. Montane forest. 2100–2150 m.

Distribution: SL—*Gereau 5130*

Asplenium dregeanum Kunze; Beentje, F.T.E.A. (Aspleniaceae): 43. 2008.

Epiphyte. Montane forest. 1500–1790 m.

Distribution: NL, RM—*Gereau 4170*

Asplenium friesiorum C. Chr.; Beentje, F.T.E.A. (Aspleniaceae): 27. 2008.

Terrestrial or epiphytic herb. Montane forest. 1950–2450 m.

Distribution: MR, RM—*Gereau 4588*

Asplenium hypomelas Kuhn; Beentje, F.T.E.A. (Aspleniaceae): 48. 2008.

Terrestrial, epiphytic, or epilithic herb. Montane forest. 2040–2200 m.

Distribution: MR—*Mwasumbi 16315*

Asplenium monanthes L.; Beentje, F.T.E.A. (Aspleniaceae): 16. 2008.

Terrestrial or epiphytic herb. Montane forest. 2200–2460 m.

Distribution: KP, NL, RM—*Gereau 4116*

Asplenium protensum Schrad.; Beentje, F.T.E.A. (Aspleniaceae): 41. 2008.

Epiphyte and on wet riverbanks. Montane forest. 1800–2380 m.

Distribution: KP, MR, RM, SL—*Gereau 4589*

Asplenium theciferum (Kunth) Mett.; Beentje, F.T.E.A. (Aspleniaceae): 55. 2008.

Epiphytic or epilithic herb. Montane forest, tree clumps in grassland. 1800–2380 m.

Distribution: KP, MR, RM, SL—*Gereau 3728*

CYATHEACEAE

1 genus and 4 species

Cyathea dregei Kunze; Edwards, F.T.E.A. (Cyatheaceae): 8. 2005.

Tree fern. Riverine forest in montane grassland. 1810 m.

Distribution: SL—*Harder 1506*

Cyathea manniana Hook.; Edwards, F.T.E.A. (Cyatheaceae): 9. 2005.

Tree fern. Montane forest. 1880–2100 m.

Distribution: MR, NL, RM—*Gereau 4227*

Cyathea mossambicensis Baker; Edwards, F.T.E.A. (Cyatheaceae): 7. 2005.

Tree fern. Montane forest. 1560–1710 m.

Distribution: RM—*Mwasumbi 16293*

Cyathea thomsonii Baker; Edwards, F.T.E.A. (Cyatheaceae): 12. 2005.

Tree fern. Montane forest. 1940 m.

Distribution: RM—*WCS SHCP 6*

DRYOPTERIDACEAE

5 genera and 8 species

Athyrium schimperi Moug. ex Fée; Verdcourt, F.T.E.A. (Woodsiaceae): 8. 2003.

Terrestrial herb. Stream banks in grassland. 2560–2570 m.

Distribution: KP, MP—*Gereau 3166*

Dryoathyrium boryanum (Willd.) Ching; Verdcourt, F.T.E.A. (Woodsiaceae): 2. 2003.

Terrestrial herb. Montane forest, riverine forest. 1250–2460 m.

Distribution: MP, NL—*Gereau 4121*

Dryopteris athamantica (Kunze) Kuntze; Roux, F.T.E.A. (Dryopteridaceae): 38. 2007.

Terrestrial herb. Montane grassland. 2080–2140 m.

Distribution: SL—*Suleiman 144*

Dryopteris pentheri (Krasser) C. Chr.; Roux, F.T.E.A. (Dryopteridaceae): 4. 2007.

Terrestrial herb. Montane forest and woodland, riverine forest. 1250–2350 m.

Distribution: MP, RM, SL—*Suleiman 100*

Dryopteris sp.

Terrestrial herb. Thicket edges in montane grassland. 2600–2650 m.

Distribution: KP—*Gereau 3198*

Polystichum transvaalense N. C. Anthony; Roux, F.T.E.A. (Dryopteridaceae): 32. 2007.

Terrestrial herb. Montane forest. 2150 m.

Distribution: MR—*Gereau 4619*

Polystichum wilsonii H. Christ; Roux, F.T.E.A. (Dryopteridaceae): 30. 2007.

Terrestrial herb. Montane forest. 2460 m.

Distribution: NL—*Gereau 4119*

Tectaria gemmifera (Fée) Alston; Shaffer-Fehre, F.T.E.A. (Dryopteridaceae): 4. 2007.

Terrestrial herb. Riverine forest. 1250 m.

Distribution: MP—*Kayombo 2349*

EQUISETACEAE

1 genus and 1 species

Equisetum ramosissimum Desf.; Verdcourt, F.T.E.A. (Equisetaceae): 1. 1999.

Terrestrial herb. Roadsides in miombo woodland, riverine forest. 1150–1500 m.

Distribution: MP, SL—*Kayombo 2459*

GLEICHENIACEAE

1 genus and 1 species

Sticherus flagellaris (Bory ex Willd.) Ching subsp. *tomentosus* (Reimers) Verdc.; Verdcourt, F.T.E.A. (Gleicheniaceae): 5. 2000.

Climbing terrestrial herb. Cliff face in montane forest. 2070 m.

Distribution: NL—*Bracebridge 76*

HYMENOPHYLLACEAE

1 genus and 1 species

Hymenophyllum capillare Desv.; Beentje, F.T.E.A. (Hymenophyllaceae): 3. 2008.

Epiphyte. Riverine forest. 1710–1740 m.

Distribution: RM—*Mwasumbi 16335*

LOMARIOPSIDACEAE

1 genus and 2 species

Elaphoglossum aubertii (Desv.) T. Moore; Mickel, F.T.E.A. (Lomariopsidaceae): 12. 2002.

Terrestrial herb. Forest patches in upland moor. 2620–2770 m.

Distribution: RM—*Mwasumbi 16265*

Elaphoglossum macropodium (Fée) T. Moore; Mickel, F.T.E.A. (Lomariopsidaceae): 18. 2002.

Epiphyte. Montane forest. 1750 m.

Distributon: RM—*Kindeketa 2983*

LYCOPODIACEAE

2 genera and 3 species

Huperzia dacrydioides (Baker) Pic. Serm.; Verdcourt, F.T.E.A. (Lycopodiaceae): 5. 2005.

Epiphyte. Montane forest. 1960–1980 m.

Distribution: MR—*Mwasumbi 16368*

Huperzia ophioglossoides (Lam.) Rothm.; Verdcourt, F.T.E.A. (Lycopodiaceae): 7. 2005.

Epiphyte. Montane forest. 2640 m.

Distribution: NL—*Gereau 4635*

Lycopodium clavatum L. subsp. *clavatum*; Verdcourt, F.T.E.A. (Lycopodiaceae): 11. 2005.

Epiphyte. Montane forest.

Distribution: RM—*Ndangalasi 1013*

NEPHROLEPIDACEAE

1 genus and 1 species

Nephrolepis undulata (Afzel. ex Sw.) J. Sm.; Verdcourt, F.T.E.A. (Oleandraceae): 5. 2001.

Terrestrial herb. Montane grassland, riverine forest, miombo woodland, cultivated fields. 1500–2440 m.

Distribution: MP, MR, RM, SL—*Suleiman 186*

OLEANDRACEAE

2 genera and 2 species

Arthropteris monocarpa (Cordem.) C. Chr.; Verdcourt, F.T.E.A. (Oleandraceae): 10. 2001.

Epiphyte. Montane forest. 1790–2400 m.

Distribution: NL, RM, SL—*Gereau 4342*

Oleandra distenta Kunze; Verdcourt, F.T.E.A. (Oleandraceae): 13. 2001.

Epiphyte. Montane forest. 1790 m.

Distribution: NL—*Gereau 4172*

OPHIOGLOSSACEAE

1 genus and 2 species

Ophioglossum reticulatum L.; Burrows, F.T.E.A. (Ophioglossaceae): 15. 2001.

Terrestrial herb. Riverine forest. 1850 m.

Distribution: SL—*Gereau 4050*

Ophioglossum vulgatum L.; Burrows, F.T.E.A. (Ophioglossaceae): 15. 2001.

Terrestrial herb. Montane grassland. 2570–2590 m.

Distribution: KP—*Gereau 3083*

OSMUNDACEAE

1 genus and 1 species

Osmunda regalis L.; Verdcourt, F.T.E.A. (Osmundaceae): 1. 1999.

Terrestrial herb. Stream banks in montane grassland. 2560 m.

Distribution: KP—*Gereau 3179*

POLYPODIACEAE

4 genera and 4 species

Drynaria volkensii Hieron.; Verdcourt, F.T.E.A. (Polypodiaceae): 12. 2001.

Epiphyte. Montane forest, riverine forest. 1850–2140 m.

Distribution: SL—*Suleiman 131*

Lepisorus excavatus (Bory ex Willd.) Ching; Verdcourt, F.T.E.A. (Polypodiaceae): 17. 2001.

Epiphytic or epilithic herb. Tree clumps in grassland, montane forest, miombo woodland. 1800–2380 m.

Distribution: KP, MP, MR, SL—*Suleiman 125A*

Loxogramme abyssinica (Baker) M. G. Price; Verdcourt, F.T.E.A. (Polypodiaceae): 33. 2001.

Epiphyte. Montane forest, riverine forest. 980–2150 m.

Distribution: NL, RM, SL—*Gereau 5156*

Pleopeltis macrocarpa (Bory ex Willd.) Kaulf.; Verdcourt, F.T.E.A. (Polypodiaceae): 26. 2001.

Epiphyte. Montane forest. 2080–2140 m.

Distribution: SL—*Suleiman 125*

PTERIDACEAE

4 genera and 5 species

Adiantum capillus-veneris L.; Verdcourt, F.T.E.A. (Adiantaceae): 61. 2002.

Terrestrial herb. Riverine forest. 1150 m.

Distribution: MP—*Kayombo 2440*

Adiantum poiretii Wikstr.; Verdcourt, F.T.E.A. (Adiantaceae): 62. 2002.

Terrestrial herb. Wet riverbanks in montane forest, riverine forest in miombo. 1275–2380 m.

Distribution: KP, MP—*Kayombo 2360*

Cheilanthes farinosa (Forssk.) Kaulf.; Verdcourt, F.T.E.A. (Adiantaceae): 28. 2002.

Terrestrial herb. Riverine forest. 1275–1300 m.

Distribution: MP, MR—*Kayombo 2359*

Pellaea pectiniformis Baker; Verdcourt, F.T.E.A. (Adiantaceae): 9. 2002.

Epilithic herb. Miombo woodland. 1780 m.

Distribution: SL—*Gereau 5101*

Pteris catoptera Kunze var. *catoptera*; Verdcourt, F.T.E.A. (Pteridaceae): 25. 2002.

Terrestrial herb. Montane forest, riverine forest. 1250–2350 m.

Distribution: MP, RM—*Kayombo 2347*

SCHIZAEACEAE

1 genus and 2 species

Mohria lepigera (Baker) Baker; Verdcourt, F.T.E.A. (Schizaeaceae): 8. 2000.

Terrestrial herb. Stream banks in montane grassland. 2560–2570 m.

Distribution: KP—*Gereau 3167*

Mohria vestita Baker; Verdcourt, F.T.E.A. (Schizaeaceae): 7. 2000.

Terrestrial herb. Rocky montane grassland. 2488–2770 m.

Distribution: RM, SL—*Suleiman 179*

SELAGINELLACEAE

1 genus and 3 species

Selaginella eublepharis A. Braun; Verdcourt, F.T.E.A. (Selaginellaceae): 10. 2005.

Terrestrial herb or epiphyte. Miombo woodland. 620 m.

Distribution: SL—*Harder 1518*

Selaginella goudotiana Spring var. *abyssinica* (Spring) Bizzarri; Verdcourt, F.T.E.A. (Selaginellaceae): 19. 2005.

Terrestrial herb. Montane forest. 1560–2350 m.

Distribution: RM, SL—*Mwasumbi 16297*

Selaginella kraussiana (Kunze) A. Braun; Verdcourt, F.T.E.A. (Selaginellaceae): 18. 2005.

Terrestrial herb. Montane forest, bamboo forest, montane grassland. 1750–2760 m.

Distribution: KP, RM—*Abeid 2897*

THELYPTERIDACEAE

3 genera and 4 species

Amauropelta bergiana (Schltdl.) Holttum var. *bergiana*; Verdcourt, F.T.E.A. (Thelypteridaceae): 6. 2006.

Terrestrial herb. Montane forest. 2460 m.

Distribution: NL—*Gereau 4120*

Christella chaseana (Schelpe) Holttum; Verdcourt, F.T.E.A. (Thelypteridaceae): 17. 2006.

Epilithic herb. Riverine forest. 980 m.

Distribution: NL—*Gereau 5163*

Christella dentata (Forssk.) Brownsey & Jermy; Verdcourt, F.T.E.A. (Thelypteridaceae): 13. 2006.

Epilithic herb. Riverine forest. 980–1400 m.

Distribution: MP, NL—*Gereau 5164*

Pneumatopteris unita (Kunze) Holttum; Verdcourt, F.T.E.A. (Thelypteridaceae): 32. 2006.

Terrestrial herb. Montane forest. 1880 m.

Distribution: NL—*Gereau 4226*

PINOPHYTA

CUPRESSACEAE

1 genus and 1 species

Juniperus procera Hochst. ex Endl.; Melville, F.T.E.A. (Gymnospermae): 16. 1958.

Tree. Montane forest, stream banks in montane grassland. 2300–2700 m.

Distribution: KP—*Congdon 20*

PODOCARPACEAE

2 genera and 2 species

Afrocarpus falcatus (Thunb.) C. N. Page; Lovett et al., 2006: 9.

Tree. Montane forest with bamboo. 2630 m.

Distribution: RM—*Abeid 2994*

Podocarpus latifolius (Thunb.) R. Br. ex Mirb.; Lovett et al., 2006: 10.

Tree. Montane forest. 1450–2380 m.

Distribution: KP, RM—*Abeid 2821*

MAGNOLIOPHYTA

ACANTHACEAE

20 genera and 52 species

Acanthopale confertiflora (Lindau) C. B. Clarke; Vollesen, F.T.E.A. (Acanthaceae 1): 217. 2008.

Shrub. Montane forest and clearings. 1800–2230 m.

Distribution: MR, NL, RM, SL—*Gereau 4423*

Acanthus ueleensis De Wild.; Vollesen, F.T.E.A. (Acanthaceae 1): 94. 2008.

Terrestrial herb. Montane and riverine forest. 1250–2040 m.

Distribution: MP, MR, RM, SL—*Kayombo 2345*

Asystasia albiflora Ensermu; Ensermu & Vollesen, F.T.E.A. (Acanthaceae 2): 455. 2010.

Terrestrial herb. Riverine forest and thickets, woodland. 1200–1250 m.

Distribution: MP—*Kayombo 2346*

Asystasia gangetica (L.) T. Anderson subsp. *micrantha* (Nees) Ensermu; Ensermu & Vollesen, F.T.E.A. (Acanthaceae 2): 459. 2010.

Scrambling terrestrial herb. Riverine forest and bushland. 474–1700 m.

Distribution: MP, NL, RM, SL—*Mwasumbi 16554*

Barleria lactiflora Brummitt & Seyani; Darbyshire, F.T.E.A. (Acanthaceae 2): 392. 2010.

Terrestrial herb. Miombo woodland. 1200–1210 m.

Distribution: MP, SL—*Gereau 4297*

Blepharis grandis C. B. Clarke; Vollesen, F.T.E.A. (Acanthaceae 1): 120. 2008.

Terrestrial herb. Grassland, woodland, riverine forest. 1250–1400 m.

Distribution: MP—*Kayombo 1061*

Blepharis katangensis De Wild.; Vollesen, F.T.E.A. (Acanthaceae 1): 125. 2008.

Terrestrial herb. Grassland. 1300–1400 m.

Distribution: MP—*Kayombo 1047*

Blepharis longifolia Lindau; Vollesen, F.T.E.A. (Acanthaceae 1): 116. 2008.

Terrestrial herb. Shrubland. 1250 m.

Distribution: MP—*J. Lovett 4581*

Brachystephanus coeruleus S. Moore subsp. *apiculatus* Champl.; Darbyshire, F.T.E.A. (Acanthaceae 2): 472. 2010.

Herbaceous climber. Montane forest. 2000 m.

Distribution: MP—*P. Lovett 445*

Brillantaisia cicatricosa Lindau; Vollesen, F.T.E.A. (Acanthaceae 1): 159. 2008.

Terrestrial herb. Riverine forest. 1250–1400 m.

Distribution: MP—*Kayombo 2351*

Dicliptera brevispicata I. Darbysh.; Darbyshire, F.T.E.A. (Acanthaceae 2): 714. 2010.

Terrestrial herb. Woodland, coffee plantations. 1300–1500 m.

Distribution: MP—*J. Lovett 4803*

Dicliptera capitata Milne-Redh.; Darbyshire, F.T.E.A. (Acanthaceae 2): 707. 2010.

Terrestrial herb. Miombo woodland. 1400 m.

Distribution: MP—*P. Lovett 97*

Dicliptera carvalhoi Lindau subsp. *nemorum* (Milne-Redh.) I. Darbysh.; Darbyshire, F.T.E.A. (Acanthaceae 2): 710. 2010.

Herbaceous climber. Riverine forest. 1150 m.

Distribution: MP—*Kayombo 2453*

Dicliptera maculata Nees subsp. *maculata*; Darbyshire, F.T.E.A. (Acanthaceae 2): 701. 2010.

Shrub. Montane forest. 2110 m.

Distribution: MR—*Gereau 4546*

Duosperma livingstoniense Vollesen; Vollesen, F.T.E.A. (Acanthaceae 1): 254. 2008.

Terrestrial herb. Miombo woodland. 1530 m.

Distribution: SL—*Gereau 4350* (type: DSM, EA, K, MO, NHT, PRE)

Duosperma tanzaniense Vollesen; Vollesen, F.T.E.A. (Acanthaceae 1): 253. 2008.

Terrestrial herb. Woodland, riverine forest. 1150–1400 m.

Distribution: MP, MR—*Bidgood 2105* (type: BR, C, CAS, DSM, K, NHT, UPS), *Kayombo 2448*

Dyschoriste albiflora Lindau; Vollesen, F.T.E.A. (Acanthaceae 1): 188. 2008.

Shrub. Riverine forest. 1200 m.

Distribution: MP, MR—*J. Lovett 4409*

Dyschoriste nyassica Gilli; Vollesen, F.T.E.A. (Acanthaceae 1): 191. 2008.

Terrestrial herb or shrub. Riverine forest. 525 m.

Distribution: SL—*Gilli 522* (type: K, W)

Dyschoriste trichocalyx (Oliv.) Lindau subsp. *verticillaris* (C. B. Clarke) Vollesen; Vollesen, F.T.E.A. (Acanthaceae 1): 192. 2008.

Terrestrial herb. Shrubland. 1300 m.

Distribution: MP—*J. Lovett 4806*

Hygrophila schulli (Buch.-Ham.) M. R. Almeida & S. M. Almeida; Vollesen, F.T.E.A. (Acanthaceae 1): 176. 2008.

Terrestrial herb. Woodland, grassland, seasonal swamps. 1250–1400 m.

Distribution: MP—*Kayombo 984*

Hypoestes aristata (Vahl) Sol. ex Roem. & Schult.; Darbyshire, F.T.E.A. (Acanthaceae 2): 724. 2010.

Terrestrial herb or shrub. Montane forest, forest edges, riverine forest. 1350–2400 m.

Distribution: MP, MR, RM—*Mwasumbi 16208*

Hypoestes forskaolii (Vahl) R. Br. subsp. *forskaolii*; Darbyshire, F.T.E.A. (Acanthaceae 2): 726. 2010.

Terrestrial herb or shrub. Montane forest, woodland. 1100–2110 m.

Distribution: MP, MR, NL—*Mwasumbi 16312*

Hypoestes triflora (Forssk.) Roem. & Schult.; Darbyshire, F.T.E.A. (Acanthaceae 2): 720. 2010.

Terrestrial herb. Montane forest, riverine forest. 1250–1980 m.

Distribution: MP, MR, RM—*Kayombo 2467*

Isoglossa bracteosa Mildbr.; Darbyshire, F.T.E.A. (Acanthaceae 2): 638. 2010.

Terrestrial herb. Montane forest. 2280–2460 m.

Distribution: NL, RM—*Mwasumbi 16236*

Isoglossa grandiflora C. B. Clarke; Darbyshire, F.T.E.A. (Acanthaceae 2): 645. 2010.

Herb or shrub. Riverine forest. 1150–1400 m.

Distribution: MP—*Kayombo 2442*

Isoglossa gregoryi (S. Moore) Lindau; Darbyshire, F.T.E.A. (Acanthaceae 2): 621. 2010.

Terrestrial herb. Disturbed montane forest. 1980 m.

Distribution: MR—*Gereau 4624*

Isoglossa multinervis I. Darbysh.; Darbyshire, F.T.E.A. (Acanthaceae 2): 635. 2010.

Shrub. Montane forest. 2500 m.

Distribution: MR—*Kayombo 1105* (type: K, MO), *Aleljung 536*

Isoglossa strigulosa C. B. Clarke; Darbyshire, F.T.E.A. (Acanthaceae 2): 628. 2010.

Terrestrial herb, often climber or straggler. Montane forest, bamboo forest. 2000–2400 m.

Distribution: KP, MP, MR, RM—*Mwasumbi 16513*

Justicia acutifolia Hedrén; Vollesen, F.T.E.A. (Acanthaceae 2): 591. 2010.

Terrestrial herb. Rocky grassland. 2100–2320 m.

Distribution: KP, SL—*Richards 14014* (type: BR, K), *Gereau 3990*

Justicia betonica L.; Vollesen, F.T.E.A. (Acanthaceae 2): 549. 2010.

Terrestrial herb. Riverine forest. 1210–1500 m.

Distribution: MP, SL—*Gereau 4390*

Justicia flava (Forssk.) Vahl; Vollesen, F.T.E.A. (Acanthaceae 2): 532. 2010.

Terrestrial herb. Riverine forest. 1210–1250 m.

Distribution: SL—*Gereau 4256*

Justicia linearispica C. B. Clarke; Vollesen, F.T.E.A. (Acanthaceae 2): 539. 2010.

Terrestrial herb. Woodland. 1720 m.

Distribution: MP—*J. Lovett 3505*

Justicia mollugo C. B. Clarke; Vollesen, F.T.E.A. (Acanthaceae 2): 589. 2010.

Terrestrial herb. Riverine forest, grassland. 1850–1940 m.

Distribution: NL, SL—*Gereau 4677*

Justicia nuttii C. B. Clarke; Vollesen, F.T.E.A. (Acanthaceae 2): 561. 2010.

Terrestrial herb. Grassland. 2360–2640 m.

Distribution: KP, MR, NL—*Suleiman 38*

Justicia nyassana Lindau; Vollesen, F.T.E.A. (Acanthaceae 2): 539. 2010.

Terrestrial herb. Riverine forest. 1400–1600 m.

Distribution: MP—*Kayombo 2477*

Justicia phyllostachys C. B. Clarke; Vollesen, F.T.E.A. (Acanthaceae 2): 586. 2010.

Terrestrial herb. Miombo woodland. 1700 m.

Distribution: NL—*Bidgood 152*

Justicia striata (Klotzsch) Bullock; Vollesen, F.T.E.A. (Acanthaceae 2): 584. 2010.

Terrestrial herb. Riverine forest, miombo woodland. 850–1800 m.

Distribution: MP, SL—*Gereau 4280*

Justicia striolata Mildbr.; Vollesen, F.T.E.A. (Acanthaceae 2): 576. 2010.

Terrestrial herb. Riverine forest, montane forest. 1350–1710 m.

Distribution: MP, RM—*Mwasumbi 16273*

Justicia sp.

Terrestrial herb. Montane forest edges. 2300 m.

Distribution: SL—*Iwarsson 1212*

Note: In *Justicia* sect. *Harnieria*, fide Hedrén (1984 annotation, NHT).

Lepidagathis eriocephala Lindau; Darbyshire, F.T.E.A. (Acanthaceae 2): 320. 2010.

Terrestrial herb. Woodland, abandoned cultivation. 1300–1400 m.

Distribution: MP—*Goetze 1083* (type: BM, BR), *Kayombo 992*

Lepidagathis lanatoglabra C. B. Clarke; Darbyshire, F.T.E.A. (Acanthaceae 2): 321. 2010.

Terrestrial herb. Grassland, woodland, abandoned cultivation. 1200–1400 m.

Distribution: MP—*Kayombo 2417*

Mimulopsis solmsii Schweinf.; Vollesen, F.T.E.A. (Acanthaceae 1): 222. 2008.

Terrestrial herb. Montane forest. 2500 m.

Distribution: MR—*Kayombo 1103*

Nelsonia canescens (Lam.) Spreng.; Vollesen, F.T.E.A. (Acanthaceae 1): 18. 2008.

Straggling terrestrial herb. Riverine forest, woodland, lawns near houses. 1300–1620 m.

Distribution: MP—*J. Lovett 4753*

Ruspolia seticalyx (C. B. Clarke) Milne-Redh.; Vollesen, F.T.E.A. (Acanthaceae 2): 486. 2010.

Suffrutex. Riverine forest. 1275–1300 m.

Distribution: MP—*Kayombo 2361*

Sclerochiton obtusisepalus C. B. Clarke; Vollesen, F.T.E.A. (Acanthaceae 1): 83. 2008.

Shrub. Montane forest. 1800 m.

Distribution: RM—*Kibure 1394*

Thunbergia alata Bojer ex Sims; Vollesen, F.T.E.A. (Acanthaceae 1): 56. 2008.

Trailing terrestrial herb or vine. Woodland, grassland, riverine forest, abandoned cultivation. 1250–2600 m.

Distribution: KP, MP, MR, SL—*Suleiman 114*

Thunbergia gregoryi S. Moore; Vollesen, F.T.E.A. (Acanthaceae 1): 52. 2008.

Trailing terrestrial herb. Roadsides. 1600–1900 m.

Distribution: MP, NL—*Iwarsson 1143*

Thunbergia kirkiana T. Anderson; Vollesen, F.T.E.A. (Acanthaceae 1): 43. 2008.

Terrestrial herb. Woodland. 1300–1600 m.

Distribution: MP—*J. Lovett 3614*

Thunbergia lancifolia T. Anderson; Vollesen, F.T.E.A. (Acanthaceae 1): 70. 2008.

Terrestrial herb. Woodland, roadsides, cultivated fields. 1250–2000 m.

Distribution: MP—*J. Lovett 5005*

Thunbergia lathyroides Burkill; Vollesen, F.T.E.A. (Acanthaceae 1): 68. 2008.

Terrestrial herb. Woodland. 1400 m.

Distribution: MP—*J. Lovett 3292A*

Thunbergia oblongifolia Oliv.; Vollesen, F.T.E.A. (Acanthaceae 1): 73. 2008.

Terrestrial herb. Bamboo forest. 2340 m.

Distribution: KP—*WCS SHCP 354*

Thunbergia petersiana Lindau; Vollesen, F.T.E.A. (Acanthaceae 1): 63. 2008.

Terrestrial herb. Riverine forest. 1160 m.

Distribution: SL—*Gereau 3696*

AIZOACEAE

1 genus and 1 species

Psammotropha myriantha Sond.; Jeffrey, F.T.E.A. (Aizoaceae): 11. 1961.

Terrestrial herb. Rocky outcrops in montane grassland. 2600–2700 m.

Distribution: KP—*Gereau 2827*

ALANGIACEAE

1 genus and 1 species

Alangium chinense (Lour.) Harms; Verdcourt, F.T.E.A. (Alangiaceae): 3. 1958.

Tree. Montane forest edges and clearings. 1840–1860 m.

Distribution: RM—*Abeid 2767*

AMARANTHACEAE

6 genera and 7 species

Achyranthes aspera L. var. *pubescens* (Moq.) C. C. Towns.; Townsend, F.T.E.A. (Amaranthaceae): 102. 1985.

Terrestrial herb. Montane forest, riverine forest. 1400–2000 m.

Distribution: MP—*J. Lovett 4678*

Achyranthes aspera L. var. *sicula* L.; Townsend, F.T.E.A. (Amaranthaceae): 104. 1985.

Terrestrial herb. Abandoned cultivation, roadsides, trailsides in montane forest. 1250–2460 m.

Distribution: MP, MR, NL—*Gereau 4521*

Achyropsis laniceps C. B. Clarke; Townsend, F.T.E.A. (Amaranthaceae): 108. 1985.

Terrestrial herb. Miombo woodland. 1400–1800 m.

Distribution: MP—*P. Lovett 111*

Aerva lanata (L.) Juss. ex Schult.; Townsend, F.T.E.A. (Amaranthaceae): 85. 1985.

Straggling herb. Riverine forest. 1210 m.

Distribution: SL—*Gereau 4267*

Celosia trigyna L.; Townsend, F.T.E.A. (Amaranthaceae): 12. 1985.

Terrestrial herb. Abandoned cultivation, woodland. 1250 m.

Distribution: MP—*J. Lovett 4456*

Cyathula cylindrica Moq.; Townsend, F.T.E.A. (Amaranthaceae): 64. 1985.

Herbaceous climber. Montane forest. 2000–2200 m.

Distribution: MP, MR—*Gereau 4522*

Cyathula polycephala Baker; Townsend, F.T.E.A. (Amaranthaceae): 61. 1985.

Scrambling herb. Montane forest edges. 2360 m.

Distribution: MR—*Gereau 3760*

Pandiaka rubrolutea (Lopr.) C. C. Towns.; Townsend, F.T.E.A. (Amaranthaceae): 115.

1985.

Terrestrial herb. Riverine forest. 1350 m.

Distribution: MP—*J. Lovett 4653*

AMARYLLIDACEAE

3 genera and 4 species

Boophone disticha (L. f.) Herb.; Nordal, F.T.E.A. (Amaryllidaceae): 21. 1982.

Terrestrial herb. Woodland. 1300 m.

Distribution: MP—*J. Lovett 4913*

Cyrtanthus breviflorus Harv.; Nordal, F.T.E.A. (Amaryllidaceae): 26. 1982.

Terrestrial herb. Montane grassland and heathland. 2600–2730 m.

Distribution: KP, MP, RM—*Abeid 2998*

Scadoxus multiflorus (Martyn) Raf. subsp. *multiflorus*; Nordal, F.T.E.A. (Amaryllidaceae): 5. 1982.

Terrestrial herb. Woodland, riverine forest, montane forest, edges of cultivation. 1250–2260 m.

Distribution: MP, SL—*Gereau 3533*

Scadoxus puniceus (L.) Friis & Nordal; Nordal, F.T.E.A. (Amaryllidaceae): 5. 1982.

Terrestrial herb. Open woodland, cultivated fields. 1400–1700 m.

Distribution: MP—*J. Lovett 3355*

Note: *J. Lovett 3351*, with narrow involucral bracts, is a possible hybrid with *S. multiflorus* subsp. *multiflorus*.

ANACARDIACEAE

5 genera and 12 species

Lannea katangensis Van der Veken; Kokwaro, F.T.E.A. (Anacardiaceae): 16. 1986.

Shrublet. Woodland. 1200–1400 m.

Distribution: MP—*Kayombo 1020*

Lannea schimperi (Hochst. ex A. Rich.) Engl.; Kokwaro, F.T.E.A. (Anacardiaceae): 19. 1986.

Tree. Miombo woodland. 1350 m.

Distribution: MP—*Kayombo 1092*

Ozoroa insignis Delile subsp. *reticulata* (Baker f.) J. B. Gillett; Kokwaro, F.T.E.A. (Anacardiaceae): 5. 1986.

Tree. Rocky gorges, lakeshore, miombo woodland. 474–1390 m.

Distribution: MP, SL—*Kayombo 2412*

Rhus anchietae Hiern vel sp. aff.; Kokwaro, F.T.E.A. (Anacardiaceae): 35. 1986.

Shrub. Montane forest. 1740 m.

Distribution: RM—*Abeid 2930*

Rhus brenanii Kokwaro; Kokwaro, F.T.E.A. (Anacardiaceae): 38. 1986.

Shrub, scandent shrub, or liana. Montane forest and forest edges, miombo woodland. 1400–2600 m.

Distribution: MP, MR, RM, SL—*Mwasumbi 16242*

Rhus glaucescens A. Rich.; Moffett, 1993: 76.

Shrub. Riverine forest. 1500 m.

Distribution: SL—*Gereau 4787*

Rhus longipes Engl. var. *elgonensis* Kokwaro; Kokwaro, F.T.E.A. (Anacardiaceae): 38. 1986.

Shrub. Riverine forest, grassland, montane forest edges. 1810–2340 m.

Distribution: RM, SL—*Harder 1474*

Rhus longipes Engl. var. *longipes*; Kokwaro, F.T.E.A. (Anacardiaceae): 37. 1986.

Shrub. Miombo woodland, lakeshores, wet meadows, disturbed montane forest. 474–1790 m.

Distribution: MP, NL, RM—*Harder 1517*

Rhus longipes Engl. var. *schinoides* R. Fern.; Kokwaro, F.T.E.A. (Anacardiaceae): 38. 1986.

Shrub or small tree. Riverine forest, thickets. 1810–2140 m.

Distribution: SL—*Suleiman 95*

Rhus ochracea Meikle; Kokwaro, F.T.E.A. (Anacardiaceae): 35. 1986.

Shrub. Lower montane forest. 2015 m.

Distribution: MR—*Gereau 7022*

Rhus pyroides Burch. var. *gracilis* (Engl.) Burtt Davy; Kokwaro, F.T.E.A. (Anacardiaceae): 32. 1986.

Shrub or small tree. Montane forest, bamboo forest, heathland. 1800–2890 m.

Distribution: KP, MP, NL, RM—*Abeid 2997*

Rhus pyroides Burch. var. *pyroides*; Moffett, 1993: 43.

Shrub or small tree. Thickets, wooded grassland, miombo woodland, disturbed montane forest, riverbanks, lakeshores. 474–2250 m.

Distribution: MP, RM, SL—*Kayombo 2369*

Rhus quartiniana A. Rich.; Kokwaro, F.T.E.A. (Anacardiaceae): 33. 1986.

Tree. Disturbed montane forest. 1790 m.

Distribution: RM—*Abeid 2906*

Sorindeia madagascariensis Thouars ex DC.; Kokwaro, F.T.E.A. (Anacardiaceae): 46. 1986.

Tree. Riverine forest, lakeshore. 474–480 m.

Distribution: NL, SL—*Gereau 3656*

Trichoscypha ulugurensis Mildbr. subsp. *submontana* (Van der Veken) Kokwaro; Kokwaro F.T.E.A. (Anacardiaceae): 51. 1986.

Tree. Montane forest. 1750 m.

Distribution: NL—*Bracebridge 107*

Trichoscypha ulugurensis Mildbr. subsp. *ulugurensis*; Kokwaro, F.T.E.A. (Anacardiaceae): 51. 1986.

Tree. Montane forest. 1830 m.

Distribution: RM—*Kibure 1408*

ANNONACEAE

5 genera and 6 species

Annona senegalensis Pers. subsp. *senegalensis*; Verdcourt, F.T.E.A. (Annonaceae): 115. 1971.

Shrub or small tree. Rocky lakeshore, woodland, abandoned cultivation. 474–1320 m.

Distribution: MP, SL—*Mwasumbi 16560*

Artabotrys monteiroae Oliv.; Verdcourt, F.T.E.A. (Annonaceae): 62. 1971.

Liana or shrub. Woodland, disturbed riverine forest. 1100–1200 m.

Distribution: MP—*Kayombo 2455*

Artabotrys stolzii Diels; Verdcourt, F.T.E.A. (Annonaceae): 64. 1971.

Liana. Riverine forest, montane forest. 1210–1880 m.

Distribution: MP, NL, SL—*Gereau 4252*

Friesodielsia obovata (Benth.) Verdc.; Verdcourt, F.T.E.A. (Annonaceae): 86. 1971.

Tree or shrub, sometimes lianescent. Riverine forest, woodland. 580–1200 m.

Distribution: MP, SL—*Kayombo 2460*

Hexalobus monopetalus (A. Rich.) Engl. & Diels var. *obovatus* Brenan; Verdcourt, F.T.E.A. (Annonaceae): 48. 1971.

Tree. Miombo woodland. 1200 m.

Distribution: MP—*J. Lovett 4040*

Uvaria lucida Bojer ex Benth. subsp. *lucida*; Verdcourt, F.T.E.A. (Annonaceae): 18. 1971.

Liana. Lakeshore shrubland, riverine forest. 474–840 m.

Distribution: SL—*Mwasumbi 16592*

ANTHERICACEAE

1 genus and 16 species

Chlorophytum affine Baker var. *affine*; Nordal et al., F.T.E.A. (Anthericaceae): 24. 1997.

Terrestrial herb. Rocky mountain slopes and grassland. 2740–2800 m.

Distribution: KP—*J. Lovett 3670*

Chlorophytum blepharophyllum Schweinf. ex Baker; Nordal et al., F.T.E.A. (Anthericaceae): 51. 1997.

Terrestrial herb. Grassland, woodland, abandoned cultivation. 1250–1600 m.

Distribution: MP—*J. Lovett 3531*

Chlorophytum cameronii (Baker) Kativu var. *cameronii*; Nordal et al., F.T.E.A. (Anthericaceae): 31. 1997.

Terrestrial herb. Rocky woodland, riverine thickets. 1200–2000 m.

Distribution: MP—*J. Lovett 3477*

Chlorophytum cameronii (Baker) Kativu var. *pterocaulon* (Welw. ex Baker) Nordal; Nordal et al., F.T.E.A. (Anthericaceae): 32. 1997.

Terrestrial herb. Woodland, montane grassland, rocky mountain slopes. 1200–2800 m.

Distribution: KP, MP—*Goldblatt 8192A*

Chlorophytum collinum (Poelln.) Nordal; Nordal et al., F.T.E.A. (Anthericaceae): 23. 1997.

Terrestrial herb. Rocky mountain slopes. 2800 m.

Distribution: KP—*Goldblatt 8192*

Chlorophytum colubrinum (Baker) Engl.; Nordal et al., F.T.E.A. (Anthericaceae): 19. 1997.

Terrestrial herb. Woodland. 1200–1250 m.

Distribution: MP—*J. Lovett 4253*

Chlorophytum comosum (Thunb.) Jacques; Nordal et al., F.T.E.A. (Anthericaceae): 55. 1997.

Terrestrial herb. Riverine forest, montane grassland. 1210–2320 m.

Distribution: MP, SL—*Gereau 4421*

Chlorophytum floribundum Baker; Nordal et al., F.T.E.A. (Anthericaceae): 47. 1997.

Terrestrial herb. Riverine forest, abandoned cultivation. 580–1250 m.

Distribution: MP, SL—*Gereau 3823*

Chlorophytum gallabatense Schweinf. ex Baker; Nordal et al., F.T.E.A. (Anthericaceae): 45. 1997.

Terrestrial herb. Riverine forest and thickets, woodland. 580–1310 m.

Distribution: MP, SL—*Gereau 3824*

Chlorophytum macrophyllum (A. Rich.) Asch.; Nordal et al., F.T.E.A. (Anthericaceae): 49. 1997.

Terrestrial herb. Stream banks in montane grassland. 2560 m.

Distribution: KP—*Gereau 3183*

Chlorophytum nubicum (Baker) Kativu; Nordal et al., F.T.E.A. (Anthericaceae): 15. 1997.

Terrestrial herb. Woodland. 1200 m.

Distribution: MP—*J. Lovett 3959*

Chlorophytum nyasae (Rendle) Kativu; Nordal et al., F.T.E.A. (Anthericaceae): 13. 1997.

Terrestrial herb. Montane grassland. 2120 m.

Distribution: SL—*Gereau 4319*

Chlorophytum rubribracteatum (De Wild.) Kativu vel sp. aff.; Nordal et al., F.T.E.A. (Anthericaceae): 36. 1997.

Terrestrial herb. Woodland. 1250–1400 m.

Distribution: MP—*J. Lovett 3650*

Chlorophytum sphacelatum (Baker) Kativu var. *milanjianum* (Rendle) Nordal; Nordal et al., F.T.E.A. (Anthericaceae): 35. 1997.

Terrestrial herb. Miombo woodland. 1200 m.

Distribution: MP—*J. Lovett 4060*

Chlorophytum stolzii (K. Krause) Kativu; Nordal et al., F.T.E.A. (Anthericaceae): 15. 1997.

Terrestrial herb. Cultivated fields, grassland. 1100–1400 m.

Distribution: MP—*Kayombo 994*

Chlorophytum subpetiolatum (Baker) Kativu; Nordal et al., F.T.E.A. (Anthericaceae): 36. 1997.

Terrestrial herb. Woodland. 1300–1950 m.

Distribution: MP—*J. Lovett 3787*

Chlorophytum zingiberastrum Nordal & A. D. Poulsen; Nordal et al., F.T.E.A. (Anthericaceae): 48. 1997.

Terrestrial herb. Woodland, riverine forest. 1200 m.

Distribution: MP—*J. Lovett 4301*

APIACEAE

13 genera and 28 species

Agrocharis incognita (C. Norman) Heywood & Jury; Townsend, F.T.E.A. (Umbelliferae): 30. 1989.

Terrestrial herb. Agricultural fields. 2200 m.

Distribution: MR—*Gereau 4511*

Agrocharis melanantha Hochst.; Townsend, F.T.E.A. (Umbelliferae): 33. 1989.

Terrestrial herb. Wet montane grassland. 2660 m.

Distribution: KP—*Abeid 2788*

Agrocharis pedunculata (Baker f.) Heywood & Jury; Townsend, F.T.E.A. (Umbelliferae): 32. 1989.

Terrestrial herb. Montane grassland. 1860–2590 m.

Distribution: KP, MP, SL—*Gereau 3870*

Alepidea peduncularis Steud. ex A. Rich.; Townsend, F.T.E.A. (Umbelliferae): 20. 1989.

Terrestrial herb. Wooded grassland, montane grassland. 1700–2700 m.

Distribution: KP, MP, MR, NL—*Suleiman 18*

Anthriscus sylvestris (L.) Hoffm. var. *sylvestris*; Townsend, F.T.E.A. (Umbelliferae): 26. 1989.

Terrestrial herb. Roadsides. 2300 m.

Distribution: MR—*Macha 78*

Centella asiatica (L.) Urb.; Townsend, F.T.E.A. (Umbelliferae): 15. 1989.

Terrestrial herb. Gardens, ditch banks. 1740 m.

Distribution: MP—*Batty 756*

Cryptotaenia africana (Hook. f.) Drude; Townsend, F.T.E.A. (Umbelliferae): 54. 1989.

Terrestrial herb. Montane forest. 1790–2110 m.

Distribution: MR, NL—*Mwasumbi 16198*

Heracleum abyssinicum (Boiss.) C. Norman; Townsend, F.T.E.A. (Umbelliferae): 121. 1989.

Terrestrial herb. Thickets and stream banks in montane grassland. 2550–2650 m.

Distribution: KP—*Gereau 3197*

Heteromorpha arborescens Cham. & Schltdl. var. *abyssinica* (Hochst. ex A. Rich.) H. Wolff; Winter & Wyk, 1996: 227.

Tree or shrub. Montane forest, riverine forest, woodland. 1250–2420 m.

Distribution: KP, MP, NL, RM, SL—*Gereau 4796*

Heteromorpha involucrata Conrath; Townsend, F.T.E.A. (Umbelliferae): 40. 1989.

Terrestrial herb. Miombo woodland, wooded grassland, riverine forest, montane grassland. 1200–2600 m.

Distribution: KP, MP, MR—*Gereau 7038*

Lefebvrea grantii S. Droop; Townsend, F.T.E.A. (Umbelliferae): 113. 1989.

Terrestrial herb. Riverine forest, woodland. 1250 m.

Distribution: MP—*J. Lovett 4512*

Peucedanum claessensii C. Norman; Townsend, F.T.E.A. (Umbelliferae): 103. 1989.

Terrestrial herb. Montane forest, thickets in montane grassland, bushland. 1430–2560 m.

Distribution: KP, MR, RM, SL—*Gereau 4748*

Peucedanum dispersum C. C. Towns.; Townsend, F.T.E.A. (Umbelliferae): 101. 1989.

Terrestrial herb. Montane grassland. 2360 m.

Distribution: KP, MR—*Hooper 1758* (type: K), *Gereau 3776*

Peucedanum cf. *kerstenii* Engl.; Townsend, F.T.E.A. (Umbelliferae): 102. 1989.

Terrestrial herb. Montane grassland. 1810 m.

Distribution: SL—*Harder 1495*

Peucedanum lynesii C. Norman; Townsend, F.T.E.A. (Umbelliferae): 96. 1989.

Terrestrial herb. Montane grassland and shrubland. 2350–2500 m.

Distribution: KP—*Brummitt 18129*

Peucedanum sp.

Terrestrial herb. Montane grassland. 2300 m.

Distribution: SL—*Iwarsson 1210*

Pimpinella aff. *alismatifolia* C. C. Towns.; Townsend, F.T.E.A. (Umbelliferae): 61. 1989.

Terrestrial herb. Rocky thickets, rocky grassland. 1800–2570 m.

Distribution: KP, NL—*Gereau 3199*

Pimpinella buchananii H. Wolff; Townsend, F.T.E.A. (Umbelliferae): 63. 1989.

Terrestrial herb. Woodland, montane grassland. 1200–2500 m.

Distribution: MP, NL, SL—*Gereau 4382*

Pimpinella caffra (Eckl. & Zeyh.) D. Dietr. subsp. *conopodioides* C. C. Towns.; Townsend, F.T.E.A. (Umbelliferae): 65. 1989.

Terrestrial herb. Rocky montane grassland. 2590 m.

Distribution: KP, MP, RM—*Hooper 1755* (type: K), *Abeid 3052*

Pimpinella duridentata C. C. Towns.; Townsend, F.T.E.A. (Umbelliferae): 61. 1989.

Terrestrial herb. Montane grassland. 2700 m.

Distribution: NL—*Gereau 4085*

Pimpinella kyimbilaensis H. Wolff; Townsend, F.T.E.A. (Umbelliferae): 65. 1989.

Terrestrial herb. Thickets in montane grassland. 2550 m.

Distribution: KP—*Gereau 3150*

Pimpinella ledermannii H. Wolff subsp. *engleriana* (H. Wolff) C. C. Towns.; Townsend, F.T.E.A. (Umbelliferae): 58. 1989.

Terrestrial herb. Miombo woodland. 1780 m.

Distribution: SL—*Gereau 5104*

Pimpinella rigidistyla C. C. Towns.; Townsend, F.T.E.A. (Umbelliferae): 64. 1989.

Terrestrial herb. 2610 m.

Distribution: MP—*Kerfoot 1743* (type: EA)

Pimpinella cf. *rigidistyla* C. C. Towns.; Townsend, F.T.E.A. (Umbelliferae): 64. 1989.

Terrestrial herb. Riverine forest, rocky grassland. 1850–2320 m.

Distribution: SL—*Gereau 4041*

Sanicula elata Buch.-Ham. ex D. Don; Townsend, F.T.E.A. (Umbelliferae): 17. 1989.

Terrestrial herb. Montane forest. 2360–2460 m.

Distribution: MP, MR, NL—*Gereau 4113*

Sium repandum Welw. ex Hiern; Townsend, F.T.E.A. (Umbelliferae): 73. 1989.

Terrestrial herb. Stream banks.

Distribution: RM—*Richards 14271*

Steganotaenia araliacea Hochst.; Townsend, F.T.E.A. (Umbelliferae): 115. 1989.

Tree. Riverine forest, montane forest edges. 1740–2070 m.

Distribution: MR, RM—*St. Clair-Thompson 738*

Steganotaenia hockii (C. Norman) C. Norman; Townsend, F.T.E.A. (Umbelliferae): 117. 1989.

Terrestrial herb. Miombo woodland, grassland. 1400–2250 m.

Distribution: MP—*J. Lovett 4868*

APOCYNACEAE

8 genera and 10 species

Carissa spinarum L.; Omino, F.T.E.A. (Apocynaceae 1): 12. 2002.

Lianescent shrub. Riverine forest, miombo woodland. 980–1530 m.

Distribution: MP, NL, SL—*Gereau 4291*

Carvalhoa campanulata K. Schum.; Omino, F.T.E.A. (Apocynaceae 1): 54. 2002.

Shrub. Montane forest. 1450–1980 m.

Distribution: MR, RM—*Mwasumbi 16349*

Diplorhynchus condylocarpon (Müll. Arg.) Pichon; Omino, F.T.E.A. (Apocynaceae 1): 56. 2002.

Tree. Woodland. 1200–1400 m.

Distribution: MP—*Kayombo 2408*

Holarrhena pubescens Wall. ex G. Don; Omino, F.T.E.A. (Apocynaceae 1): 92. 2002.

Shrub. Lakeshore forest and bushland. 474–480 m.

Distribution: SL—*Gereau 3650*

Landolphia buchananii (Hallier f.) Stapf; Omino, F.T.E.A. (Apocynaceae 1): 19. 2002.

Liana or shrub. Riverine forest, montane forest. 840–1950 m.

Distribution: MP, RM, SL—*Gereau 3687*

Rauvolfia caffra Sond.; Omino, F.T.E.A. (Apocynaceae 1): 61. 2002.

Tree. Riverine forest, pond margins in montane grassland. 1290–1840 m.

Distribution: MP, SL—*Gereau 3922*

Rauvolfia mannii Stapf; Omino, F.T.E.A. (Apocynaceae 1): 63. 2002.

Shrub. Montane forest. 1750 m.

Distribution: RM—*Kindeketa 2976*

Saba comorensis (Bojer ex A. DC.) Pichon; Omino, F.T.E.A. (Apocynaceae 1): 34. 2002.

Liana. Lakeshore shrubland. 474–480 m.

Distribution: SL—*Mwasumbi 16603*

Tabernaemontana stapfiana Britten; Omino, F.T.E.A. (Apocynaceae 1): 49. 2002.

Tree. Riverine forest. 2000 m.

Distribution: MP—*J. Lovett 4990*

Tabernaemontana ventricosa Hochst. ex A. DC.; Omino, F.T.E.A. (Apocynaceae 1): 49. 2002.

Tree. Riverine forest. 890–1160 m.

Distribution: SL—*Gereau 3698*

AQUIFOLIACEAE

1 genus and 1 species

Ilex mitis (L.) Radlk. var. *mitis*; Verdcourt, F.T.E.A. (Aquifoliaceae): 3. 1968.

Tree or shrub. Montane forest. 1450–2340 m.

Distribution: KP, MR, NL, RM, SL—*Mwasumbi 16303*

ARACEAE

2 genera and 2 species

Amorphophallus abyssinicus (A. Rich.) N. E. Br. subsp. *unyikae* (Engl. & Gehrm.) Ittenb. ex Govaerts & Frodin; Ittenbach, 2003: 81.

Terrestrial herb. Edges of cultivation, woodland. 1250–1300 m.

Distribution: MP—*J. Lovett 3692*

Culcasia falcifolia Engl.; Mayo, F.T.E.A. (Araceae): 18. 1985.

Epiphyte. Montane forest. 1750 m.

Distribution: RM—*Abeid 2898*

ARALIACEAE

3 genera and 6 species

Cussonia arborea Hochst. ex A. Rich.; Bamps,

1974: 104.

Tree. Montane woodland, miombo woodland. 910–2000 m.

Distribution: MP, RM, SL—*Gereau 4069*

Cussonia spicata Thunb.; Bamps, 1974: 115.

Tree. Montane forest, forest/woodland transition. 1620–2100 m.

Distribution: RM, SL—*Gereau 4060*

Polyscias fulva (Hiern) Harms; Bamps, 1974: 122.

Tree. Riverine forest, montane forest. 1350–1900 m.

Distribution: MP, RM—*Kayombo 1076*

Schefflera lukwangulensis (Tennant) Bernardi; Bamps, 1974: 132.

Tree. Montane forest. 1610 m.

Distribution: RM—*Kibure 1399*

Schefflera myriantha (Baker) Drake; Bamps, 1974: 133.

Liana or tree. Montane forest. 1730–2015 m.

Distribution: MR, NL—*Gereau 4145*

Schefflera stolzii Harms; Bamps, 1974: 135.

Liana. Montane forest. 1740–2360 m.

Distribution: MR, RM, SL—*Stolz 2044* (type: BM, K, MO, P, WAG), *Abeid 2929*

ARECACEAE

1 genus and 1 species

Phoenix reclinata Jacq.; Dransfield, F.T.E.A. (Palmae): 15. 1986.

Tree. Montane grassland and forest edges, riverine forest. 890–2120 m.

Distribution: SL—*Gereau 4344*

ARISTOLOCHIACEAE

1 genus and 1 species

Aristolochia hockii De Wild.; Verdcourt, F.T.E.A. (Aristolochiaceae): 7. 1986.

Terrestrial herb. Miombo woodland, abandoned cultivation. 1200–1250 m.

Distribution: MP—*J. Lovett 4212*

ASCLEPIADACEAE

18 genera and 38 species

Asclepias randii S. Moore; Brown, F.T.A. 4(1): 351. 1902.

Terrestrial herb. Montane grassland. 2150 m.

Distribution: SL—*Gereau 5139*

Asclepias sp.

Terrestrial herb. Montane grassland. 2800 m.

Distribution: KP—*Goldblatt 8215*

Aspidoglossum angustissimum (K. Schum.) Bullock; Bullock, 1952: 418.

Terrestrial herb. Wooded grassland. 1250 m.

Distribution: MP—*J. Lovett 4221*

Brachystelma kituloense Goyder; Goyder, 1990: 729.

Terrestrial herb. Montane grassland. 2250–2800 m.

Distribution: KP—*Brummitt 18092* (type: DSM, EA, K, MO, PRE), *Richards 18501*

Ceropegia claviloba Werderm.; Huber, 1957: 153.

Twining vine. Montane forest. 1700–1880 m.

Distribution: NL, RM—*Gereau 4224*

Ceropegia meyeri-johannis Engl.; Huber, 1957: 154.

Twining vine. Montane forest. 2380 m.

Distribution: KP—*Gereau 3129*

Ceropegia papillata N. E. Br. var. *cordiloba* (Werderm.) H. Huber; Huber, 1957: 152.

Twining vine. Montane forest/woodland transition. 2140 m.

Distribution: SL—*Gereau 4002*

Ceropegia racemosa N. E. Br.; Huber, 1957: 97.

Terrestrial herb. Miombo woodland. 1200 m.

Distribution: MP—*J. Lovett 3969*

Ceropegia stenantha K. Schum.; Huber, 1957: 123.

Terrestrial herb. Rocky montane grassland. 2700 m.

Distribution: KP—*Gereau 2837*

Cryptolepis oblongifolia (Meisn.) Schltr.; Brown, F.T.A. 4(1): 249. 1902.

Shrub. Woodland, riverine forest. 1200–1400 m.

Distribution: MP—*Lovett 4054*

Cynanchum rungweense Bullock; Liede, 1996a: 335.

Herbaceous climber. Montane forest, tree clumps in grassland. 2200–2380 m.

Distribution: KP, MR—*Gereau 3724*

Dregea abyssinica K. Schum.; Bullock, F.W.T.A. 2: 97. 1963.

Liana. Riverine forest. 1300–1400 m.

Distribution: MP—*Kayombo 1025*

Glossostelma carsonii (N. E. Br.) Bullock; Bullock, 1952: 415.

Terrestrial herb. Miombo woodland. 1780–1790 m.

Distribution: SL—*Gereau 3876*

Gomphocarpus physocarpus E. Mey.; Bullock, F.W.T.A. 2: 92. 1963.

Terrestrial herb. 2870 m.

Distribution: KP—*WCS SHCP 373*

Gongronema angolense (N. E. Br.) Bullock; Bullock, F.W.T.A. 2: 98. 1963.

Herbaceous climber.

Distribution: MR—*Stolz 548*

Pachycarpus concolor E. Mey.; Smith, 1988: 411.

Terrestrial herb. Montane grassland. 2550 m.

Distribution: KP—*Goldblatt 8219*

Pachycarpus distinctus (N. E. Br.) Bullock; Bullock, 1953b: 335.

Terrestrial herb. Disturbed woodland. 1910 m.

Distribution: SL—*Gereau 4073*

Pachycarpus goetzei (K. Schum.) Bullock; Bullock, 1953b: 338.

Terrestrial herb. Miombo woodland. 1700 m.

Distribution: NL—*Bidgood 154*

Pachycarpus lineolatus (Decne.) Bullock; Bullock, 1953b: 333.

Terrestrial herb. Miombo woodland. 1200–1600 m.

Distribution: MP—*J. Lovett 5038*

Pachycarpus spurius (N. E. Br.) Bullock; Bullock, 1953b: 338.

Terrestrial herb. Woodland, abandoned cultivation. 1200–1250 m.

Distribution: MP—*J. Lovett 4028*

Periploca linearifolia Quart.-Dill. & A. Rich.; Venter, 2003: 103.

Liana. Montane forest, forest/woodland transition. 1800–2140 m.

Distribution: RM, SL—*Suleiman 104*

Raphionacme aff. *lanceolata* Schinz; Brown, F.T.A. 4(1): 274. 1902.

Terrestrial herb. Miombo woodland. 1200–1250 m.

Distribution: MP—*J. Lovett 3972*

Raphionacme longituba E. A. Bruce; Bruce, 1937: 419.

Terrestrial herb. Miombo woodland. 1200 m.

Distribution: MP—*J. Lovett 3881*

Raphionacme welwitschii Schltr. & Rendle; Brown, F.T.A. 4(1): 273. 1902.

Terrestrial herb. Grassland, woodland. 1200–1600 m.

Distribution: MP—*Gereau 3493*

Schizoglossum alpestre K. Schum.; Brown, F.T.A. 4(1): 364. 1902.

Terrestrial herb. Montane grassland. 2800 m.

Distribution: KP—*Goldblatt 8197*

Schizoglossum sp.

Terrestrial herb. Montane grassland. 2800 m.

Distribution: KP—*Goldblatt 8195*

Stathmostelma gigantiflorum K. Schum.; Bullock, 1953a: 52.

Terrestrial herb. Woodland. 1250 m.

Distribution: MP—*J. Lovett 3879*

Stathmostelma spectabile (N. E. Br.) Schltr.; Bullock, 1953a: 56.

Terrestrial herb. Riverine forest edge. 1160 m.

Distribution: SL—*Gereau 3697*

Tacazzea apiculata Oliv.; Venter, 2003: 106.

Liana. Montane forest edges. 2300 m.

Distribution: MR, NL—*Magogo 2400*

Tacazzea conferta N. E. Br.; Venter, 2003: 106.

Liana. Montane forest edges. 2280–2650 m.

Distribution: MP, MR, NL—*Iwarsson 1159*

Tylophora aff. *apiculata* K. Schum.; Brown, F.T.A. 4(1): 407. 1903.

Herbaceous climber. Montane forest. 1850 m.

Distribution: RM—*Bidgood 82*

Tylophora conspicua N. E. Br.; Brown, F.T.A. 4(1): 405. 1903.

Herbaceous climber. Riverine forest. 1320 m.

Distribution: MP—*Gereau 6122*

Tylophora erubescens (Liede & Meve) Liede; Liede, 1996b: 206.

Herbaceous climber. Montane forest. 1920 m.

Distribution: NL—*Gereau 4212*

Xysmalobium kaessneri S. Moore; Moore, 1908: 295.

Terrestrial herb. Miombo woodland. 1200 m.

Distribution: MP—*J. Lovett 3944*

Xysmalobium reticulatum N. E. Br.; Brown, F.T.A. 4(2): 303. 1902.

Terrestrial herb. Miombo woodland. 1790 m.

Distribution: SL—*Gereau 3909*

Xysmalobium undulatum (L.) W. T. Aiton; Goyder, 2003: 131.

Terrestrial herb. Montane grassland. 2080–2140 m.

Distribution: SL—*Suleiman 115*

Xysmalobium sp.

Terrestrial herb. Riverbanks in montane grassland. 2560 m.

Distribution: KP—*Gereau 3178*

Xysmalobium sp.

Terrestrial herb. Thin soil among boulders. 2600–2800 m.

Distribution: KP, MP—*Gereau 3502*

ASPARAGACEAE

1 genus and 7 species

Asparagus africanus Lam. var. *puberulus* (Baker) Sebsebe; Sebsebe, F.T.E.A. (Asparagaceae): 8. 2006.

Shrub. Edges of cultivation. 1300 m.

Distribution: MP—*J. Lovett 3548*

Asparagus asparagoides (L.) Druce; Sebsebe, F.T.E.A. (Asparagaceae): 19. 2006.

Liana. Montane forest. 2300 m.

Distribution: MR—*Gereau 4581*

Asparagus buchananii Baker; Sebsebe, F.T.E.A. (Asparagaceae): 14. 2006.

Shrub. Miombo woodland. 1300–2140 m.

Distribution: MP, RM, SL—*J. Lovett 4809*

Asparagus flagellaris (Kunth) Baker; Sebsebe, F.T.E.A. (Asparagaceae): 10. 2006.

Terrestrial herb. Abandoned cultivation, miombo woodland. 1250–1400 m.

Distribution: MP—*J. Lovett 3750*

Asparagus laricinus Burch.; Sebsebe, F.T.E.A. (Asparagaceae): 5. 2006.

Shrub. Montane bushland, scree slopes. 1980–2300 m.

Distribution: MR, SL—*Mwasumbi 16545*

Asparagus setaceus (Kunth) Jessop; Sebsebe, F.T.E.A. (Asparagaceae): 9. 2006.

Liana. Montane forest and thickets, riverine forest. 1500–2230 m.

Distribution: NL, RM, SL—*Harder 1504*

Asparagus virgatus Baker; Sebsebe, F.T.E.A. (Asparagaceae): 4. 2006.

Shrub. Scree slopes. 2060–2120 m.

Distribution: SL—*Gereau 4445*

ASPHODELACEAE

4 genera and 8 species

Aloe leedalii S. Carter; Carter, F.T.E.A. (Aloaceae): 9. 1994.

Terrestrial herb. Rocky slopes. 2950 m.

Distribution: KP—*Leedal 5503*

Aloe nuttii Baker; Carter, F.T.E.A. (Aloaceae): 9. 1994.

Terrestrial herb. Rocky slopes, montane grassland. 2500–2800 m.

Distribution: KP, MP—*P. Lovett 93*

Bulbine abyssinica A. Rich.; Whitehouse, F.T.E.A. (Asphodelaceae): 11. 2002.

Terrestrial herb. Woodland, montane grassland, rocky ridges. 1200–2850 m.

Distribution: KP, MP, SL—*Abeid 2792*

Kniphofia goetzei Engl.; Whitehouse, F.T.E.A. (Asphodelaceae): 4. 2002.

Terrestrial herb. Montane forest. 1980–2500 m.

Distribution: MR, NL, SL—*Gereau 4629*

Kniphofia grantii Baker; Whitehouse, F.T.E.A. (Asphodelaceae): 7. 2002.

Terrestrial herb. Woodland, montane grassland. 1400–2500 m.

Distribution: KP, MP—*J. Lovett 5040*

Kniphofia paludosa Engl.; Whitehouse, F.T.E.A. (Asphodelaceae): 9. 2002.

Terrestrial or aquatic herb. Wet places in montane grassland, in running water of streams, swamps. 2300–2770 m.

Distribution: KP—*Goetze 1278* (type: B [destroyed]), *Gereau 3182*

Kniphofia princeae (A. Berger) Marais; Whitehouse, F.T.E.A. (Asphodelaceae): 6. 2002.

Terrestrial herb. Montane grassland. 2500 m.

Distribution: MP—*P. Lovett 89*

Trachyandra saltii (Baker) Oberm.; Whitehouse, F.T.E.A. (Asphodelaceae): 16. 2002.

Terrestrial herb. Montane grassland. 2150–2800 m.

Distribution: KP, SL—*Gereau 5140*

ASTERACEAE

55 genera and 139 species

Acanthospermum hispidum DC.; Beentje & Hind, F.T.E.A. (Compositae 3): 766. 2005.

Terrestrial herb. Disturbed montane forest.

Distribution: RM—*WCS SHCP 233*

Acmella caulirhiza Delile; Beentje & Hind, F.T.E.A. (Compositae 3): 730. 2005.

Terrestrial herb. Shallow ponds in montane grassland, riverine forest, edges of cultivation. 1200–2080 m.

Distribution: MP, RM, SL—*Gereau 3917*

Adenostemma caffrum DC.; Smith, F.T.E.A. (Compositae 3): 821. 2005.

Terrestrial herb. 2080–2140 m.

Distribution: SL—*Suleiman 124*

Adenostemma mauritianum DC.; Smith, F.T.E.A. (Compositae 3): 824. 2005.

Terrestrial herb. Montane forest, *Eucalyptus* plantations, riverine brushland. 1660–2300 m.

Distribution: MR, RM, SL—*Abeid 2886*

Ageratinastrum polyphyllum (Baker) Mattf.; Jeffrey & Beentje, F.T.E.A. (Compositae 1): 276. 2000.

Terrestrial herb. Montane grassland. 2488 m.

Distribution: SL—*Suleiman 167*

Ageratum conyzoides L.; Smith, F.T.E.A. (Compositae 3): 827. 2005.

Terrestrial herb. Open woodland, edges of cultivation. 1250–1400 m.

Distribution: MP, RM—*J. Lovett 3687*

Ageratum houstonianum Mill. var. *houstonianum*; Smith, F.T.E.A. (Compositae 3): 830. 2005.

Terrestrial herb.

Distribution: RM—*Semsei 1553*

Anisopappus chinensis (L.) Hook. & Arn. subsp. *buchwaldii* (O. Hoffm.) S. Ortiz, Paiva & Rodr. Oubiña; Beentje, F.T.E.A. (Compositae 2): 344. 2002.

Terrestrial herb. Montane forest edges, grassland. 1980–2500 m.

Distribution: KP, MP, MR, NL, RM—*Gereau 4593*

Artemisia afra Jacq. ex Willd.; Beentje, F.T.E.A. (Compositae 2): 529. 2002.

Shrub or woody-based herb. Montane grassland. 1960–2840 m.

Distribution: KP, RM—*Suleiman 29*

Aspilia mossambicensis (Oliv.) Wild; Beentje & Hind, F.T.E.A. (Compositae 3): 751. 2005.

Shrub. Riverine forest edges, bamboo forest, grassland, woodland. 1200–2000 m.

Distribution: MP, NL, SL—*Kayombo 2354*

Aspilia pluriseta Schweinf.; Beentje & Hind, F.T.E.A. (Compositae 3): 749. 2005.

Terrestrial herb. Edges of cultivation, woodland. 1200–1400 m.

Distribution: MP—*J. Lovett 3847*

Aster tansaniensis W. Lippert; Beentje, F.T.E.A. (Compositae 2): 470. 2002.

Terrestrial herb. Montane grassland. 2230–2500 m.

Distribution: KP, RM, SL—*Gereau 3591*

Athrixia rosmarinifolia (Sch. Bip. ex Walp.) Oliv. & Hiern var. *rosmarinifolia*; Beentje, F.T.E.A. (Compositae 2): 456. 2002.

Terrestrial herb. Bushland. 2360 m.

Distribution: KP, MP—*Myembe 56*

Athrixia subsimplex Brenan; Beentje, F.T.E.A. (Compositae 2): 456. 2002.

Terrestrial herb. Montane grassland. 1800–2700 m.

Distribution: KP, MP, MR—*Machaga 25*

Berkheya echinacea (Harv.) O. Hoffm. ex Burtt Davy subsp. *polyacantha* (Baker) Roessler; Halliday & Wilmot-Dear, F.T.E.A. (Compositae 1): 291. 2000.

Terrestrial herb. Montane grassland. 1940–2840 m.

Distribution: KP, MP, RM, SL—*Suleiman 24*

Berkheya zeyheri Oliv. & Hiern; Halliday & Wilmot-Dear, F.T.E.A. (Compositae 1): 293. 2000.

Terrestrial herb. Rocky mountain slopes. 2250–2800 m.

Distribution: KP, MP—*Goldblatt 8202*

Bidens flagellata (Sherff) Mesfin; Mesfin, F.T.E.A. (Compositae 3): 796. 2005.

Terrestrial herb. Montane grassland. 2140–2150 m.

Distribution: SL—*Suleiman 155*

Bidens kilimandscharica (O. Hoffm.) Sherff; Mesfin, F.T.E.A. (Compositae 3): 783. 2005.

Terrestrial herb. Montane forest. 2110 m.

Distribution: MR—*Gereau 4566*

Bidens ochracea (O. Hoffm.) Sherff; Mesfin, F.T.E.A. (Compositae 3): 789. 2005.

Terrestrial herb. Rocky montane grassland. 2740 m.

Distribution: KP—*Abeid 3075*

Bidens pilosa L.; Mesfin, F.T.E.A. (Compositae 3): 806. 2005.

Terrestrial herb. Riverine forest, montane forest, woodland. 1160–1900 m.

Distribution: MP, NL, RM, SL—*Gereau 3678*

Bidens pinnatipartita (O. Hoffm.) Wild; Mesfin, F.T.E.A. (Compositae 3): 783. 2005.

Soft-wooded shrub. Montane forest edges, montane grassland. 1940–2300 m.

Distribution: MP, MR, SL—*Gereau 4657*

Bidens schimperi Sch. Bip. ex Walp.; Mesfin, F.T.E.A. (Compositae 3): 803. 2005.

Terrestrial herb. Grassland, montane forest. 1300–2400 m.

Distribution: MP, RM—*Mwasumbi 16268*

Bidens steppia (Steetz) Sherff; Mesfin, F.T.E.A. (Compositae 3): 801. 2005.

Terrestrial herb. Grassland, miombo woodland, abandoned cultivation. 1250–1800 m.

Distribution: MP—*J. Lovett 4619*

Blumea axillaris (Lam.) DC.; Beentje, F.T.E.A. (Compositae 2): 319. 2002.

Terrestrial herb. Riverine forest, woodland. 560–1250 m.

Distribution: MP, SL—*J. Lovett 3761*

Bothriocline inyangana N. E. Br. var. *amplexicaulis* (Muschl.) C. Jeffrey; Jeffrey & Beentje, F.T.E.A. (Compositae 1): 147. 2000.

Shrub. Montane forest edges, grassland, roadsides. 1950–2240 m.

Distribution: MR, RM—*Gereau 4520*

Bothriocline inyangana N. E. Br. var. *inyangana*; Jeffrey & Beentje, F.T.E.A. (Compositae 1): 147. 2000.

Shrub. Abandoned cultivation. 2000 m.

Distribution: MP—*P. Lovett 252*

Bothriocline mbalensis (Wild & G. V. Pope) C. Jeffrey; Jeffrey & Beentje, F.T.E.A. (Compositae 1): 145. 2000.

Terrestrial herb. Miombo woodland. 1550–1800 m.

Distribution: MP—*Gereau 7002*

Bothriocline pectinata (O. Hoffm.) Wild & G. V. Pope; Jeffrey & Beentje, F.T.E.A. (Compositae 1): 155. 2000.

Shrub or woody-based herb. Montane forest. 1750–2354 m.

Distribution: RM, SL—*Abeid 2925*

Brachythrix glomerata (Mattf.) C. Jeffrey; Jeffrey & Beentje, F.T.E.A. (Compositae 1): 278. 2000.

Terrestrial herb. Miombo woodland. 1200–2140 m.

Distribution: MP, SL—*Gereau 4409*

Brachythrix stolzii (S. Moore) Wild & G. V. Pope; Jeffrey & Beentje, F.T.E.A. (Compositae 1): 277. 2000.

Terrestrial herb. Montane forest edges, grassland. 1980–2525 m.

Distribution: MR, NL, RM—*Stolz 2595* (type: BM, K, MO), *Abeid 2830*

Carduus nyassanus (S. Moore) R. E. Fr. subsp.

kikuyorum (R. E. Fr.) C. Jeffrey; Jeffrey & Beentje, F.T.E.A. (Compositae 1): 49. 2000.

Terrestrial herb. Montane forest, grassland. 2400 m.

Distribution: KP, MR, RM—*Richards 6461*

Chrysanthemoides monilifera (L.) Norl. subsp. *septentrionalis* Norl.; Beentje, F.T.E.A. (Compositae 2): 535. 2002.

Shrub. Scree slopes. 2060–2120 m.

Distribution: SL—*Gereau 4438*

Cineraria deltoidea Sond.; Jeffrey & Beentje, F.T.E.A. (Compositae 3): 563. 2005.

Terrestrial herb. Montane grassland, edges of cultivation. 2200–2550 m.

Distribution: KP, MP, MR—*Gereau 3154*

Cirsium buchwaldii O. Hoffm.; Jeffrey & Beentje, F.T.E.A. (Compositae 1): 54. 2000.

Terrestrial herb. Stream banks in montane grassland. 2400–2590 m.

Distribution: KP—*Suleiman 13*

Conyza gouanii (L.) Willd.; Beentje, F.T.E.A. (Compositae 2): 493. 2002.

Terrestrial herb. Open woodland, riverine forest. 1650–1850 m.

Distribution: SL—*Gereau 4047*

Conyza limosa O. Hoffm.; Beentje, F.T.E.A. (Compositae 2): 508. 2002.

Terrestrial herb. Pond margins, stream banks, seasonally waterlogged grassland. 1250–2550 m.

Distribution: KP, MP, SL—*Gereau 3913*

Conyza pyrrhopappa Sch. Bip. ex A. Rich.; Beentje, F.T.E.A. (Compositae 2): 499. 2002.

Shrub. Woodland, riverine forest. 1300–1780 m.

Distribution: MP—*J. Lovett 4874*

Conyza stricta Willd.; Beentje, F.T.E.A. (Compositae 2): 500. 2002.

Terrestrial herb. Riverine forest. 1500 m.

Distribution: SL—*Gereau 3950*

Conyza tigrensis Oliv. & Hiern; Beentje, F.T.E.A. (Compositae 2): 497. 2002.

Terrestrial herb. Montane grassland. 2230–2650 m.

Distribution: KP, MR, SL—*Gereau 3770*

Crassocephalum crepidioides (Benth.) S. Moore; Jeffrey & Beentje, F.T.E.A. (Compositae 3): 610. 2005.

Terrestrial herb. Riverine forest, montane forest. 1210–1670 m.

Distribution: RM, SL—*Gereau 4298*

Crassocephalum montuosum (S. Moore) Milne-Redh.; Jeffrey & Beentje, F.T.E.A. (Compositae 3): 602. 2005.

Terrestrial herb. Grassland, montane forest. 1700–2100 m.

Distribution: MR, RM—*Richards 6551*

Crassocephalum rubens (Juss. ex Jacq.) S. Moore var. *rubens*; Jeffrey & Beentje, F.T.E.A. (Compositae 3): 609. 2005.

Terrestrial herb. Disturbed forest, woodland, grassland, cultivation. 1100–2100 m.

Distribution: MP, MR, RM—*J. Lovett 3782*

Crassocephalum rubens (Juss. ex Jacq.) S. Moore var. *sarcobasis* (DC.) C. Jeffrey & Beentje; Jeffrey & Beentje, F.T.E.A. (Compositae 3): 609. 2005.

Terrestrial herb. Miombo woodland. 1400 m.

Distribution: MP—*P. Lovett 110*

Crassocephalum vitellinum (Benth.) S. Moore; Jeffrey & Beentje, F.T.E.A. (Compositae 3): 606. 2005.

Terrestrial herb. Montane forest, woodland, roadsides. 1400–2300 m.

Distribution: MP, RM—*J. Lovett 4882*

Crepis newii Oliv. & Hiern subsp. *oliveriana* (Kuntze) C. Jeffrey & Beentje; Jeffrey & Beentje, F.T.E.A. (Compositae 1): 70. 2000.

Terrestrial herb. Montane grassland, miombo woodland, cultivation, roadsides. 1740–2600 m.

Distribution: KP, MP, MR, RM, SL—*Suleiman 26*

Dichrocephala chrysanthemifolia (Blume) DC. var. *chrysanthemifolia*; Beentje, F.T.E.A. (Compositae 2): 461. 2002.

Terrestrial herb. Montane grassland, montane forest edges. 2300–2640 m.

Distribution: KP, MP, NL—*Gereau 4642*

Dichrocephala integrifolia (L. f.) Kuntze subsp. *integrifolia*; Beentje, F.T.E.A. (Compositae 2): 459. 2002.

Terrestrial herb. Coffee plantation, old cultivation, riverine forest, montane forest edges.

1250–2420 m.

Distribution: MP, RM—*J. Lovett 5008*

Dicoma anomala Sond.; Jeffrey & Beentje, F.T.E.A. (Compositae 1): 14. 2000.

Terrestrial herb. Grassland, woodland. 1200–2500 m.

Distribution: KP, MP—*Kayombo 2431*

Dicoma kirkii Harv. subsp. *vaginata* (O. Hoffm.) G. V. Pope; Jeffrey & Beentje, F.T.E.A. (Compositae 1): 17. 2000.

Terrestrial herb. Wooded grassland. 980 m.

Distribution: SL—*Kibure 1484*

Echinops giganteus A. Rich.; Jeffrey & Beentje, F.T.E.A. (Compositae 1): 39. 2000.

Terrestrial herb. Montane grassland, cultivation. 1650–2500 m.

Distribution: MP, MR—*P. Lovett 494*

Echinops lanatus C. Jeffrey & Mesfin; Jeffrey & Beentje, F.T.E.A. (Compositae 1): 39. 2000.

Terrestrial herb. Montane grassland. 2400–2850 m.

Distribution: KP, MR, NL—*Bidgood 121* (type: K, NHT), *Abeid 2860*

Elephantopus scaber L. subsp. *plurisetus* (O. Hoffm.) Philipson; Jeffrey & Beentje, F.T.E.A. (Compositae 1): 280. 2000.

Terrestrial herb. Woodland, grassland, riverine forest. 1200–2040 m.

Distribution: MP, MR, SL—*Gereau 3964*

Emilia abyssinica (Sch. Bip. ex A. Rich.) C. Jeffrey var. *abyssinica*; Jeffrey & Beentje, F.T.E.A. (Compositae 3): 575. 2005.

Terrestrial herb. Miombo woodland. 1200 m.

Distribution: MP—*J. Lovett 4064*

Emilia caespitosa Oliv.; Jeffrey & Beentje, F.T.E.A. (Compositae 3): 581. 2005.

Terrestrial herb. Miombo woodland. 1475 m.

Distribution: MP—*Kayombo 2398*

Emilia coccinea (Sims) G. Don; Jeffrey & Beentje, F.T.E.A. (Compositae 3): 583. 2005.

Terrestrial herb. Abandoned cultivation, woodland. 1200–1800 m.

Distribution: MP, RM—*Kayombo 1001*

Emilia integrifolia Baker; Jeffrey & Beentje, F.T.E.A. (Compositae 3): 592. 2005.

Terrestrial herb. Grassland, montane forest. 1250–2660 m.

Distribution: KP, MP, MR—*Gereau 3107*

Emilia tenera (O. Hoffm.) C. Jeffrey; Jeffrey & Beentje, F.T.E.A. (Compositae 3): 592. 2005.

Terrestrial herb. Montane grassland. 1800–2740 m.

Distribution: KP, RM, SL—*J. Lovett 774*

Erythrocephalum longifolium Benth. ex Oliv.; Jeffrey & Beentje, F.T.E.A. (Compositae 1): 27. 2000.

Terrestrial herb. Wooded grassland, miombo woodland, abandoned cultivation. 1250–1900 m.

Distribution: MP, MR—*Gereau 7050*

Felicia boehmii O. Hoffm.; Beentje, F.T.E.A. (Compositae 2): 475. 2002.

Terrestrial herb. Miombo woodland. 1400–1475 m.

Distribution: MP—*Kayombo 2397*

Galinsoga parviflora Cav.; Beentje & Hind, F.T.E.A. (Compositae 3): 763. 2005.

Terrestrial herb. Riverine forest, edges of cultivation, weed of bean fields. 1250–1500 m.

Distribution: MP, SL—*Gereau 3961*

Gerbera piloselloides (L.) Cass.; Jeffrey & Beentje, F.T.E.A. (Compositae 1): 12. 2000.

Terrestrial herb. Montane forest remnants, woodland. 1320–1840 m.

Distribution: MP, SL—*Gereau 3549*

Gerbera viridifolia (DC.) Sch. Bip. subsp. *viridifolia*; Jeffrey & Beentje, F.T.E.A. (Compositae 1): 10. 2000.

Terrestrial herb. Montane grassland. 1810–2700 m.

Distribution: KP, NL, SL—*Taylor 8527*

Guizotia scabra (Vis.) Chiov. subsp. *scabra*; Beentje & Hind, F.T.E.A. (Compositae 3): 771. 2005.

Soft-wooded shrub. Montane bushland. 2280 m.

Distribution: MR—*Mwasumbi 16168*

Gutenbergia cordifolia Benth. ex Oliv. var. *cordifolia*; Jeffrey & Beentje, F.T.E.A. (Compositae 1): 124. 2000.

Terrestrial herb. Roadsides. 474–480 m.

Distribution: SL—*Mwasumbi 16588*

Haplocarpha thunbergii Less.; Halliday & Wilmot-Dear, F.T.E.A. (Compositae 1): 287. 2000.

Terrestrial herb. Montane grassland. 2320–2570 m.

Distribution: KP, SL—*Gereau 3997*

Helichrysum abietinum O. Hoffm.; Beentje, F.T.E.A. (Compositae 2): 417. 2002.

Subshrub. Montane grassland, heathland. 2550–2960 m.

Distribution: KP, RM—*Goetze 1151* (type: BR, K), *Abeid 3008*

Helichrysum aureum (Houtt.) Merr. var. *monocephalum* (DC.) Hilliard; Beentje, F.T.E.A. (Compositae 2): 437. 2002.

Terrestrial herb. Montane grassland. 2640 m.

Distribution: KP, NL—*Gereau 4648*

Helichrysum foetidum (L.) Moench; Beentje, F.T.E.A. (Compositae 2): 434. 2002.

Terrestrial herb. Montane grassland, montane forest/woodland transition, riverine forest. 2170–2900 m.

Distribution: KP, MR, SL—*Gereau 3999*

Helichrysum forskahlii (J. F. Gmel.) Hilliard & B. L. Burtt var. *forskahlii*; Beentje, F.T.E.A. (Compositae 2): 415. 2002.

Terrestrial herb or shrub. Heathland, montane grassland, woodland. 1400–2770 m.

Distribution: KP, MP, MR, RM, SL—*Mwasumbi 16259*

Helichrysum globosum Sch. Bip.; Beentje, F.T.E.A. (Compositae 2): 417. 2002.

Terrestrial herb. Montane grassland. 2550 m.

Distribution: KP—*Abeid 3108*

Helichrysum herbaceum (Andrews) Sweet; Beentje, F.T.E.A. (Compositae 2): 434. 2002.

Terrestrial herb. Montane grassland. 2120–2700 m.

Distribution: KP, NL, RM, SL—*Gereau 4322*

Helichrysum kalandanum Lisowski; Beentje, F.T.E.A. (Compositae 2): 427. 2002.

Terrestrial herb. Montane forest, montane grassland, woodland. 1890–2360 m.

Distribution: KP, MP, MR, NL, SL—*Gereau 4221*

Helichrysum kirkii Oliv. & Hiern; Beentje, F.T.E.A. (Compositae 2): 429. 2002.

Terrestrial herb. Montane grassland, miombo woodland. 1280–2840 m.

Distribution: KP, MP, SL—*Suleiman 19*

Helichrysum milne-redheadii Brenan vel sp. aff.; Lisowski, 1989: 166.

Subshrub. Montane grassland. 1940 m.

Distribution: SL—*Gereau 4658*

Helichrysum nitens Oliv. & Hiern; Beentje, F.T.E.A. (Compositae 2): 433. 2002.

Terrestrial herb. Montane grassland. 2500–2740 m.

Distribution: KP, MP—*Gereau 3203*

Helichrysum nudifolium (L.) Less.; Beentje, F.T.E.A. (Compositae 2): 420. 2002.

Terrestrial herb. Montane grassland. 2230–2700 m.

Distribution: NL, SL—*Gereau 3588*

Helichrysum odoratissimum (L.) Sweet; Beentje, F.T.E.A. (Compositae 2): 410. 2002.

Terrestrial herb. Montane grassland, montane shrubland. 1760–2360 m.

Distribution: MR, NL—*Gereau 3769*

Helichrysum quartinianum A. Rich.; Beentje, F.T.E.A. (Compositae 2): 438. 2002.

Terrestrial herb. Miombo woodland. 1820 m.

Distribution: SL—*Gereau 3907*

Helichrysum schimperi (Sch. Bip. ex A. Rich.) Moeser; Beentje, F.T.E.A. (Compositae 2): 418. 2002.

Straggling terrestrial herb. Montane bushland, montane forest. 1780–2240 m.

Distribution: MR, RM—*Mwasumbi 16507*

Helichrysum setosum Harv.; Beentje, F.T.E.A. (Compositae 2): 436. 2002.

Terrestrial herb. Pyrethrum fields. 2200 m.

Distribuition: MR, RM—*Gereau 4524*

Helichrysum tillandsiifolium O. Hoffm.; Beentje, F.T.E.A. (Compositae 2): 412. 2002.

Subshrub. Rocky montane grassland, woodland. 1920–2740 m.

Distribution: KP, NL—*Wiland 149A*

Hypericophyllum compositarum Steetz; Beentje & Hind, F.T.E.A. (Compositae 3): 720. 2005.

Terrestrial herb. Miombo woodland. 1600 m.

Distribution: MP—*P. Lovett 229*

Hypericophyllum elatum (O. Hoffm.) N. E. Br.; Beentje & Hind, F.T.E.A. (Compositae 3): 720. 2005.

Terrestrial herb. Riverine forest. 1200 m.

Distribution: MP—*J. Lovett 4401*

Inula mannii (Hook. f.) Oliv. & Hiern; Beentje, F.T.E.A. (Compositae 2): 326. 2002.

Terrestrial herb or suffrutex. Montane forest. 2354–2500 m.

Distribution: MR, RM—*Kayombo 1107*

Inula rungwensis Beentje; Beentje, F.T.E.A. (Compositae 2): 328. 2002.

Terrestrial herb. Montane grassland. 2600 m.

Distribution: KP—*J. Lovett 3655* (type: K, MO)

Inula shirensis Oliv.; Beentje, F.T.E.A. (Compositae 2): 323. 2002.

Terrestrial herb. Woodland, grassland. 1250–1400 m.

Distribution: MP—*Kayombo 983*

Inula stolzii Mattf.; Beentje, F.T.E.A. (Compositae 2): 326. 2002.

Terrestrial herb. Montane grassland. 2590 m.

Distribution: KP—*Abeid 3053*

Inula sp.

Terrestrial herb. Miombo woodland. 2000 m.

Distribution: MP—*P. Lovett 257*

Lactuca inermis Forssk.; Jeffrey & Beentje, F.T.E.A. (Compositae 1): 83. 2000.

Terrestrial herb. Fallow fields, woodland, montane grassland. 1250–2340 m.

Distribution: MP, MR, RM—*Mwasumbi 16528*

Laggera brevipes Oliv. & Hiern; Beentje, F.T.E.A. (Compositae 2:) 355. 2002.

Terrestrial herb. Disturbed woodland. 1200 m.

Distribution: MP—*Kayombo 2422*

Launaea cornuta (Hochst. ex Oliv. & Hiern) C. Jeffrey; Jeffrey & Beentje, F.T.E.A. (Compositae 1): 104. 2000.

Terrestrial herb. Montane forest. 1920 m.

Distribution: RM—*Ndangalasi SHCP 111*

Launaea rarifolia (Oliv. & Hiern) Boulos; Jeffrey & Beentje, F.T.E.A. (Compositae 1): 101. 2000.

Terrestrial herb. Woodland. 1400 m.

Distribution: MP—*J. Lovett 3361*

Lopholaena dolichopappa (O. Hoffm.) S. Moore; Jeffrey & Beentje, F.T.E.A. (Compositae 3): 570. 2005.

Terrestrial herb. Montane bushland. 2200–2300 m.

Distribution: MR—*Mwasumbi 16544*

Lopholaena trianthema (O. Hoffm.) B. L. Burtt; Jeffrey & Beentje, F.T.E.A. (Compositae 3): 572. 2005.

Shrub. Woodland. 1400 m.

Distribution: MP—*J. Lovett 3331*

Melanthera pungens Oliv. & Hiern var. *albinervia* (O. Hoffm.) Beentje; Beentje & Hind, F.T.E.A. (Compositae 3): 742. 2005.

Terrestrial herb. Woodland, riverine forest. 1200–1790 m.

Distribution: MP, SL—*Gereau 3902*

Microglossa pyrifolia (Lam.) Kuntze; Beentje, F.T.E.A. (Compositae 2): 482. 2002.

Shrub. Edges of cultivation. 1250 m.

Distribution: MP—*J. Lovett 3703*

Nidorella microcephala Steetz; Beentje, F.T.E.A. (Compositae 2): 490. 2002.

Terrestrial herb. Edges of cultivation. 1290 m.

Distribution: MP—*Gereau 6114*

Osteospermum monocephalum (Oliv. & Hiern) Norl.; Beentje, F.T.E.A. (Compositae 2): 532. 2002.

Terrestrial herb. Montane grassland and shrubland. 1810–2340 m.

Distribution: MP, SL—*Harder 1494*

Pleiotaxis pulcherrima Steetz; Jeffrey & Beentje, F.T.E.A. (Compositae 1): 22. 2000.

Terrestrial herb. Miombo woodland. 1400 m.

Distribution: MP—*P. Lovett 103*

Schistostephium artemisiifolium Baker; Beentje, F.T.E.A. (Compositae 2): 526. 2002.

Terrestrial herb. Montane grassland. 2500 m.

Distribution: MP—*P. Lovett 5*

Senecio doryphorus Mattf.; Jeffrey & Beentje, F.T.E.A. (Compositae 3): 634. 2005.

Terrestrial herb. Montane grassland. 2700 m.

Distribution: NL—*Gereau 4077*

Senecio erubescens Aiton; Jeffrey & Beentje, F.T.E.A. (Compositae 3): 667. 2005.

Terrestrial herb. Montane grassland, abandoned cultivation. 2600–2700 m.

Distribution: KP—*Taylor 8531*

Senecio immixtus C. Jeffrey; Jeffrey & Beentje, F.T.E.A. (Compositae 3): 640. 2005.

Terrestrial herb. Montane grassland. 2570 m.

Distribution: KP—*Gereau 3098*

Senecio inornatus DC.; Jeffrey & Beentje, F.T.E.A. (Compositae 3): 633. 2005.

Terrestrial herb. Montane grassland. 2360–2570 m.

Distribution: KP, MR—*Gereau 3782*

Senecio cf. *mesogrammoides* O. Hoffm.; Jeffrey & Beentje, F.T.E.A. (Compositae 3): 649. 2005.

Terrestrial herb. Montane grassland. 1940 m.

Distribution: SL—*Gereau 4692*

Senecio microalatus C. Jeffrey; Jeffrey & Beentje, F.T.E.A. (Compositae 3): 634. 2005.

Terrestrial herb. Stream bank in montane grassland. 2310 m.

Distribution: MP—*Milne-Redhead 10214* (type: K).

Senecio pergamentaceus Baker; Jeffrey & Beentje, F.T.E.A. (Compositae 3): 643. 2005.

Terrestrial herb. Miombo woodland. 1790 m.

Distribution: SL—*Gereau 3880*

Senicio purpureus L.; Jeffrey & Beentje, F.T.E.A. (Compositae 3): 668. 2005.

Terrestrial herb. Montane grassland. 2590 m.

Distribution: KP, RM—*Abeid 3048*

Senecio ruwenzoriensis S. Moore; Jeffrey & Beentje, F.T.E.A. (Compositae 3): 643. 2005.

Terrestrial herb. Riverine forest. 1850 m.

Distribution: SL—*Gereau 4048*

Senecio cf. *semiamplexifolius* De Wild.; Jeffrey & Beentje, F.T.E.A. (Compositae 3): 642. 2005.

Terrestrial herb. Montane grassland. 2400–2840 m.

Distribution: KP—*Abeid 2787*

Senecio aff. *subsessilis* Oliv. & Hiern; Jeffrey & Beentje, F.T.E.A. (Compositae 3): 627. 2005.

Terrestrial herb. Montane forest and grassland. 2200–2830 m.

Distribution: KP, MR—*Kayombo 1102*

Senecio syringifolius O. Hoffm.; Jeffrey & Beentje, F.T.E.A. (Compositae 3): 621. 2005.

Herbaceous climber. Montane forest. 1960–1980 m.

Distribution: MR—*Mwasumbi 16363*

Senecio urundensis S. Moore; Jeffrey & Beentje, F.T.E.A. (Compositae 3): 642. 2005.

Terrestrial herb. Woodland. 1720 m.

Distribution: MP—*J. Lovett 3506*

Solanecio mannii (Hook. f.) C. Jeffrey; Jeffrey & Beentje, F.T.E.A. (Compositae 3): 678. 2005.

Tree or shrub. Riverine forest, montane forest. 1600–2240 m.

Distribution: MP, MR, RM—*J. Lovett 4831*

Sonchus bipontini Asch.; Jeffrey & Beentje, F.T.E.A. (Compositae 1): 95. 2000.

Terrestrial herb. Montane grassland. 2100 m.

Distribution: MR—*Richards 6540*

Sonchus luxurians (R. E. Fr.) C. Jeffrey; Jeffrey & Beentje, F.T.E.A. (Compositae 1): 96. 2000.

Terrestrial herb. Woodland, wet montane grassland. 1720–2660 m.

Distribution: KP, MP—*Abeid 2790*

Sphaeranthus randii S. Moore var. *randii*; Beentje, F.T.E.A. (Compositae 2): 386. 2002.

Emergent aquatic herb. Seasonal ponds in montane grassland. 1810–1840 m.

Distribution: SL—*Gereau 3916*

Stoebe kilimandscharica O. Hoffm.; Beentje, F.T.E.A. (Compositae 2): 452. 2002.

Shrub. Disturbed montane grassland, heathland. 2500–2890 m.

Distribution: KP, RM—*Gereau 3139*

Tolpis capensis (L.) Sch. Bip.; Jeffrey & Beentje, F.T.E.A. (Compositae 1): 66. 2000.

Terrestrial herb. Miombo woodland, clearings in montane forest, roadsides in cultivation. 1790–2350 m.

Distribution: MR, RM, SL—*Gereau 3881*

Tridax procumbens L.; Beentje & Hind, F.T.E.A. (Compositae 3): 764. 2005.

Terrestrial herb. Roadsides, abandoned cultivation. 474–1250 m.

Distribution: MP, SL—*Gereau 3788*

Vernonia adoensis Sch. Bip. ex Walp.; Jeffrey & Beentje, F.T.E.A. (Compositae 1): 245. 2000.

Shrub. Woodland, thickets near cultivation, disturbed lower montane forest. 1300–1350 m.

Distribution: MP, RM—*Gereau 6096*

Vernonia alticola G. V. Pope; Jeffrey & Beentje, F.T.E.A. (Compositae 1): 205. 2000.

Terrestrial herb. Montane grassland. 2700 m.

Distribution: KP—*Taylor 8526*

Vernonia calvoana (Hook. f.) Hook. f. subsp. *leucocalyx* (O. Hoffm.) C. Jeffrey; Jeffrey & Beentje, F.T.E.A. (Compositae 1): 238. 2000.

Shrub. Disturbed montane forest, forest edges. 1980–2100 m.

Distribution: MR, RM—*Gereau 4627*

Vernonia calyculata S. Moore; Jeffrey & Beentje, F.T.E.A. (Compositae 1): 207. 2000.

Terrestrial herb. Miombo woodland. 1790 m.

Distribution: SL—*Gereau 3900*

Vernonia exsertiflora Baker var. *exsertiflora*; Jeffrey & Beentje, F.T.E.A. (Compositae 1): 181. 2000.

Shrub. Bushland, woodland, grassland. 1250–1400 m.

Distribution: MP—*J. Lovett 4580*

Vernonia glabra (Steetz) Vatke var. *glabra*; Jeffrey & Beentje, F.T.E.A. (Compositae 1): 230. 2000.

Terrestrial herb. Woodland. 1250 m.

Distribution: MP—*J. Lovett 4566*

Vernonia glabra (Steetz) Vatke var. *laxa* (Steetz) Brenan; Jeffrey & Beentje, F.T.E.A. (Compositae 1): 230. 2000.

Terrestrial herb. Grassland. 1100 m.

Distribution: MP—*Carter 2466*

Vernonia jelfiae S. Moore var. *albida* G. V. Pope; Jeffrey & Beentje, F.T.E.A. (Compositae 1): 253. 2000.

Terrestrial herb. Woodland, grassland. 1250–1300 m.

Distribution: MP—*J. Lovett 4617*

Vernonia karaguensis Oliv. & Hiern; Jeffrey & Beentje, F.T.E.A. (Compositae 1): 200. 2000.

Shrub or terrestrial herb. Montane forest edges, montane grassland, woodland. 1250–2110 m.

Distribution: MP, MR, SL—*Gereau 4691*

Vernonia meiostephana C. Jeffrey; Jeffrey & Beentje, F.T.E.A. (Compositae 1): 211. 2000.

Terrestrial herb. Old cultivation. 1250 m.

Distribution: MP—*J. Lovett 4548*

Vernonia melleri Oliv. & Hiern var. *superba* (O. Hoffm.) C. Jeffrey; Jeffrey & Beentje, F.T.E.A. (Compositae 1): 233. 2000.

Terrestrial herb. Woodland, edges of cultivation. 1250 m.

Distribution: MP—*J. Lovett 4114*

Vernonia musofensis S. Moore var. *miamensis* (S. Moore) G. V. Pope; Jeffrey & Beentje, F.T.E.A. (Compositae 1): 258. 2000.

Terrestrial herb. Miombo woodland. 1600 m.

Distribution: MP—*P. Lovett 238*

Vernonia myriantha Hook. f.; Jeffrey & Beentje, F.T.E.A. (Compositae 1): 195. 2000.

Shrub. Riverine forest. 1500–2250 m.

Distribution: MP, MR, RM—*J. Lovett 4692*

Vernonia perrottetii Sch. Bip. ex Walp.; Jeffrey & Beentje, F.T.E.A. (Compositae 1): 256. 2000.

Terrestrial herb. Grassland. 1100 m.

Distribution: MP—*Carter 2474*

Vernonia praemorsa Muschl.; Jeffrey & Beentje, F.T.E.A. (Compositae 1): 224. 2000.

Terrestrial herb. Montane grassland. 2250 m.

Distribution: MP—*J. Lovett 4848*

Vernonia pteropoda Oliv. & Hiern; Jeffrey & Beentje, F.T.E.A. (Compositae 1): 184. 2000.

Terrestrial herb. Montane forest. 1760 m.

Distribution: MR, RM—*Kibure 1380*

Vernonia suprafastigiata Klatt vel sp. aff.; Jeffrey & Beentje, F.T.E.A. (Compositae 1): 198. 2000.

Shrub. Miombo woodland. 1300 m.

Distribution: MP—*J. Lovett 4808*

Vernonia syringifolia O. Hoffm. vel sp. aff.; Jeffrey & Beentje, F.T.E.A. (Compositae 1): 190. 2000.

Straggling shrub. Montane forest edges. 2110 m.

Distribution: MR—*Gereau 4552*

Vernonia aff. *thomsoniana* Oliv. & Hiern; Jeffrey & Beentje, F.T.E.A. (Compositae 1): 192. 2000.

Shrub. Grassland. 1300 m.

Distribution: MP—*P. Lovett 381*

Vernonia tolypophora Mattf.; Jeffrey & Beentje, F.T.E.A. (Compositae 1): 241. 2000.

Shrub. Roadsides, edges of cultivation, woodland, montane grassland. 1300–2200 m.

Distribution: MP, MR, RM, SL—*Gereau 4772*

Vernonia ugandensis S. Moore; Jeffrey & Beentje, F.T.E.A. (Compositae 1): 261. 2000.

Terrestrial herb. Miombo woodland. 1530 m.

Distribution: SL—*Gereau 4375*

Vernonia usafuensis O. Hoffm.; Jeffrey & Beentje, F.T.E.A. (Compositae 1): 232. 2000.

Terrestrial herb. Montane grassland, montane forest edges. 1800–2600 m.

Distribution: MP, MR, RM—*Goetze 1039* (type: BR, K, P), *P. Lovett 492*

BALANOPHORACEAE

1 genus and 1 species

Sarcophyte sanguinea Sparrm. subsp. *piriei* (Hutch.) B. Hansen; Hansen, F.T.E.A. (Balanophoraceae): 3. 1993.

Parasitic terrestrial herb. Woodland, *Eucalyptus* plantation, roadsides in coffee plantation. 1310–1500 m.

Distribution: MP—*Schmidt 1128*

BALSAMINACEAE

1 genus and 15 species

Impatiens assurgens Baker f.; Grey-Wilson, F.T.E.A. (Balsaminaceae): 57. 1982.

Terrestrial herb. Woodland, montane grassland. 1300–2500 m.

Distribution: MP, MR, SL—*Gereau 4756*

Impatiens austrotanzanica Grey-Wilson; Grey-Wilson, F.T.E.A. (Balsaminaceae): 66. 1982.

Terrestrial herb. Montane forest, forest/woodland transition, montane grassland. 1550–2360 m.

Distribution: MR, RM, SL—*Kibure 1414*

Impatiens confusa Grey-Wilson subsp. *confusa*; Grey-Wilson, F.T.E.A. (Balsaminaceae): 45. 1982.

Terrestrial herb. Montane forest. 1750–1980 m.

Distribution: MR, NL, RM—*Jones & Pócs 6331/P* (type: K), *Mwasumbi 16359*

Impatiens cribbii (Grey-Wilson) Grey-Wilson; Grey-Wilson, F.T.E.A. (Balsaminaceae): 65. 1982.

Terrestrial herb. Near streams in montane forest, montane grassland. 2380–2590 m.

Distribution: KP—*Cribb 10764* (type: DSM, EA, K), *Goldblatt 8238*

Impatiens eryaleia Launert subsp. *eryaleia*; Grey-Wilson, F.T.E.A. (Balsaminaceae): 30. 1982.

Terrestrial herb. Montane forest. 1900–2350 m.

Distribution: NL, RM, SL—*Gereau 4206*

Impatiens eryaleia Launert subsp. *mbeyaensis* Grey-Wilson; Grey-Wilson, F.T.E.A. (Balsaminaceae): 30. 1982.

Terrestrial herb. Montane forest. 1560–1710 m.

Distribution: MR, RM—*Cribb 11312* (type: DSM, EA, K), *Mwasumbi 16285*

Impatiens flammea Gilg; Grey-Wilson, F.T.E.A. (Balsaminaceae): 71. 1982.

Terrestrial herb. Montane forest. 2460 m.

Distribution: NL, RM—*Goetze 1147* (type: B), *Gereau 4115*

Impatiens gomphophylla Baker f.; Grey-Wilson, F.T.E.A. (Balsaminaceae): 64. 1982.

Terrestrial herb. Montane grassland, tree clumps, montane forest edges, woodland. 1300–2760 m.

Distribution: KP, MP, MR, RM—*Abeid 2826*

Impatiens hochstetteri Warb. subsp. *hochstetteri*; Grey-Wilson, F.T.E.A. (Balsaminaceae): 46. 1982.

Terrestrial herb. Montane forest, heathland. 1560–2810 m.

Distribution: KP, NL, RM—*Mwasumbi 16294*

Impatiens hydrogetonoides Launert subsp. *hydrogetonoides*; Grey-Wilson, F.T.E.A. (Balsaminaceae): 39. 1982.

Terrestrial herb. Montane forest. 2000 m.

Distribution: MP—*J. Lovett 3851*

Impatiens hydrogetonoides Launert subsp. *kituloensis* Grey-Wilson; Grey-Wilson, F.T.E.A. (Balsaminaceae): 39. 1982.

Terrestrial herb. Stream banks, montane grassland, montane forest edges. 2380–2960 m.

Distribution: KP—*Cribb 10771* (type: DSM,

EA, K), *Goldblatt 8237*

Impatiens leedalii Grey-Wilson; Grey-Wilson, F.T.E.A. (Balsaminaceae): 54. 1982.

Terrestrial herb. Montane forest. 2380–2775 m.

Distribution: KP—*Cribb 10770* (type: DSM, EA, K), *Goldblatt 8263*

Impatiens polyantha Gilg; Grey-Wilson, F.T.E.A. (Balsaminaceae): 54. 1982.

Terrestrial herb. Montane grassland, roadsides. 1900–2600 m.

Distribution: MP, MR, RM, SL—*Gereau 3499*

Impatiens polyantha Gilg × *austrotanzanica* Grey-Wilson; Grey-Wilson, F.T.E.A. (Balsaminaceae): 57. 1982.

Terrestrial herb. Montane forest. 1960–2640 m.

Distribution: MR, NL—*Gereau 4634*

Impatiens polyantha Gilg × *rosulata* Grey-Wilson; Grey-Wilson, F.T.E.A. (Balsaminacaeae): 55. 1982.

Terrestrial herb. Montane grassland with tree clumps. 2360 m.

Distribution: MR—*Gereau 3743*

Impatiens pseudozombensis Grey-Wilson; Grey-Wilson, F.T.E.A. (Balsaminaceae): 36. 1982.

Terrestrial herb. Montane forest. 1920–2240 m.

Distribution: MR, SL—*Gereau 5129*

Impatiens rosulata Grey-Wilson; Grey-Wilson, F.T.E.A. (Balsaminaceae): 57. 1982.

Terrestrial herb. Montane grassland. 2400–2740 m.

Distribution: KP, NL—*Cribb 10761* (type: DSM, EA, K), *Suleiman 23*

Impatiens rubromaculata Warb. subsp. *grandiflora* Grey-Wilson; Grey-Wilson, F.T.E.A. (Balsaminaceae): 35. 1982.

Terrestrial herb. Montane forest. 2000–2354 m.

Distribution: MP, MR, NL, RM—*Richards 9774* (type: B, K), *Mwasumbi 16214*

Impatiens volkensii Warb. subsp. *volkensii*; Frimodt-Møller & Grey-Wilson, 1999: 180.

Terrestrial herb or soft-wooded shrub. Stream banks in montane forest, riverine brushland. 1660–1790 m.

Distribution: NL, RM—*Gereau 4168*

BASELLACEAE

1 genus and 1 species

Basella alba L.; Verdcourt, F.T.E.A. (Basellaceae): 2. 1968.

Twining herb. Montane forest. 1870–1980 m.

Distribution: MR, RM—*Mwasumbi 16362*

BEGONIACEAE

1 genus and 5 species

Begonia johnstonii Oliv. ex Hook. f.; Sands & Beentje, F.T.E.A. (Begoniaceae): 12. 2006.

Terrestrial herb. Montane grassland. 2140–2150 m.

Distribution: SL—*Suleiman 151*

Begonia meyeri-johannis Engl.; Plana, F.T.E.A. (Begoniaceae): 4. 2006.

Epilithic or epiphytic herb. Montane forest. 1500–2500 m.

Distribution: MR, RM, SL—*Mwasumbi 16223*

Begonia princeae Gilg; Sands & Beentje, F.T.E.A. (Begoniaceae): 14. 2006.

Terrestrial herb. Woodland, abandoned cultivation. 1200–1350 m.

Distribution: MP—*J. Lovett 5003*

Begonia aff. *rumpiensis* Kupicha; Kupicha, Fl. Zamb. 4: 502. 1978.

Terrestrial herb. Rocks in riverine forest. 1160–1320 m.

Distribution: MP, SL—*Gereau 3689*

Begonia sutherlandii Hook. f.; Sands & Beentje, F.T.E.A. (Begoniaceae): 17. 2006.

Epiphytic and epilithic herb. Montane forest and clearings, riverine forest. 1320–2380 m.

Distribution: KP, MP, RM, SL—*Gereau 3603*

BERBERIDACEAE

1 genus and 1 species

Berberis holstii Engl.; Polhill, F.T.E.A. (Berberidaceae): 1. 1966.

Shrub. Heathland. 2460 m.

Distribution: RM—*Geilinger s.n.*

BIGNONIACEAE

3 genera and 3 species

Kigelia africana (Lam.) Benth.; Bidgood et al., F.T.E.A. (Bignoniaceae): 43. 2006.

Tree. Open woodland. 1350 m.

Distribution: MP—*Lovett 4769*

Markhamia obtusifolia (Baker) Sprague; Bidgood et al., F.T.E.A. (Bignoniaceae): 35. 2006.

Tree or shrub. Woodland, edges of cultivation, riverine forest. 830–1400 m.

Distribution: MP, SL—*Kayombo 2449*

Tecomaria capensis (Thunb.) Spach subsp. *nyassae* (Oliv.) Brummitt; Bidgood et al., F.T.E.A. (Bignoniaceae): 25. 2006.

Tree or shrub. Wooded grassland, riverine forest, montane forest. 1500–2750 m.

Distribution: MP, MR, NL, RM, SL—*Suleiman 132*

BOMBACACEAE

1 genus and 1 species

Adansonia digitata L.; Beentje, F.T.E.A. (Bombacaceae): 4. 1989.

Tree. Secondary forest on lakeshore. 474 m.

Distribution: SL—*Gereau 3666*

BORAGINACEAE

7 genera and 10 species

Cordia africana Lam.; Verdcourt, F.T.E.A. (Boraginaceae): 31. 1991.

Tree. Woodland, old cultivation. 1250–1400 m.

Distribution: MP—*Kayombo 955*

Cynoglossum amplifolium Hochst ex A. DC.; Verdcourt, F.T.E.A. (Boraginaceae): 104. 1991.

Terrestrial herb. Montane grassland. 2400–2700 m.

Distribution: KP, NL, RM—*Abeid 2796*

Cynoglossum coeruleum A. DC. subsp. *johnstonii* (Baker) Verdc.; Verdcourt, F.T.E.A. (Boraginaceae): 109. 1991.

Terrestrial herb. Montane grassland, woodland. 1200–2400 m.

Distribution: KP, MP, MR, RM, SL—*Suleiman 14*

Cynoglossum lanceolatum Forssk.; Verdcourt, F.T.E.A. (Boraginaceae): 106. 1991.

Terrestrial herb. Miombo woodland. 1530 m.

Distribution: SL—*Gereau 4377*

Ehretia cymosa Thonn.; Verdcourt, F.T.E.A. (Boraginaceae): 37. 1991.

Tree or shrub. Riverine forest, montane forest. 1800–1825 m.

Distribution: MP, RM—*Kayombo 1112*

Heliotropium ovalifolium Forssk.; Verdcourt, F.T.E.A. (Boraginaceae): 75. 1991.

Terrestrial herb. Edges of cultivation. 1290 m.

Distribution: MP—*Gereau 6115*

Lithospermum afromontanum Weim.; Verdcourt, F.T.E.A. (Boraginaceae): 77. 1991.

Terrestrial herb. Montane forest edges, montane grassland and heathland. 2080–2830 m.

Distribution: KP, MP, NL, SL—*Suleiman 136*

Myosotis vestergrenii Stroh; Verdcourt, F.T.E.A. (Boraginaceae): 87. 1991.

Terrestrial herb. Montane grassland. 2750 m.

Distribution: KP, NL—*Sturtz 232*

Trichodesma physaloides (Fenzl) A. DC.; Verdcourt, F.T.E.A. (Boraginaceae): 100. 1991.

Terrestrial herb. Abandoned cultivation, grassland. 1250–2250 m.

Distribution: MP—*J. Lovett 4608*

Trichodesma zeylanicum (Burm. f.) R. Br.; Verdcourt, F.T.E.A. (Boraginaceae): 92. 1991.

Terrestrial herb. Woodland. 1200–1250 m.

Distribution: MP—*Kayombo 2418*

BRASSICACEAE

2 genera and 2 species

Capsella bursa-pastoris (L.) Medik.; Jonsell, F.T.E.A. (Cruciferae): 29. 1982.

Terrestrial herb. Cultivated fields. 2280 m.

Distribution: MR—*Mwasumbi 16162*

Cardamine africana L.; Jonsell, F.T.E.A. (Cruciferae): 38. 1982.

Terrestrial herb. Montane forest. 2320–2460

m.

Distribution: NL, SL—*Gereau 4117*

BUDDLEJACEAE

2 genera and 5 species

Buddleja axillaris Willd. ex Roem. & Schult.; Leeuwenberg, 1979: 27.

Liana. Riverine forest. 1210 m.

Distribution: SL—*Gereau 4261*

Buddleja pulchella N. E. Br. vel sp. aff.; Leeuwenberg, 1979: 68.

Tree. Montane forest. 2120 m.

Distribution: NL—*Bracebridge 73*

Buddleja salviifolia (L.) Lam.; Leeuwenberg, 1979: 77.

Shrub or tree. Montane forest. 1780–2500 m.

Distribution: KP, MP, RM—*P. Lovett 478*

Nuxia congesta R. Br. ex Fresen.; Bruce & Lewis, F.T.E.A. (Loganiaceae): 44. 1960.

Tree. Riverine forest, tree clumps in montane grassland. 2200–2360 m.

Distribution: MP, MR, RM—*Gereau 3764*

Nuxia floribunda Benth.; Bruce & Lewis, F.T.E.A. (Loganiaceae): 43. 1960.

Tree. Riverine forest, montane forest edges. 474–2350 m.

Distribution: MR, NL, RM, SL—*Mwasumbi 16609*

BURSERACEAE

1 genus and 4 species

Commiphora africana (A. Rich.) Engl. var. *rubriflora* (Engl.) Wild; Gillett, F.T.E.A. (Burseraceae): 47. 1991.

Tree. Riverine forest. 1250–1290 m.

Distribution: MP—*Gereau 6111*

Commiphora eminii Engl. subsp. *zimmermannii* (Engl.) J. B. Gillett; Gillett, F.T.E.A. (Burseraceae): 57. 1991.

Tree. Riverine forest. 1210 m.

Distribution: SL—*Gereau 4275*

Commiphora mollis (Oliv.) Engl.; Gillett, F.T.E.A. (Burseraceae): 54. 1991.

Tree. Lakeshore forest, riverine forest. 480–580 m.

Distribution: SL—*Gereau 3808*

Commiphora mossambicensis (Oliv.) Engl.; Gillett, F.T.E.A. (Burseraceae): 58. 1991.

Tree. Woodland. 1200–1400 m.

Distribution: MP—*Gereau 6116*

CAESALPINIACEAE

10 genera and 25 species

Afzelia quanzensis Welw.; Brenan, F.T.E.A. (Leguminosae–Caesalpinioideae): 125. 1967.

Tree. Lakeshore forest. 480 m.

Distribution: SL—*Gereau 3653*

Bauhinia petersiana Bolle; Brenan, F.T.E.A. (Leguminosae–Caesalpinioideae): 211. 1967.

Tree. Woodland, riverine forest edges. 1250–1400 m.

Distribution: MP—*Kayombo 968*

Bauhinia thonningii Schumach.; Wilczek, 1952: 275.

Tree. Woodland. 1200–1530 m.

Distribution: MP, SL—*Gereau 3826*

Brachystegia bussei Harms; Hoyle, F.T.E.A. (Leguminosae–Caesalpinioideae): 170. 1967.

Tree. Miombo woodland. 1300–1400 m.

Distribution: MP—*Kayombo 1030*

Brachystegia floribunda Benth.; Hoyle, F.T.E.A. (Leguminosae–Caesalpinioideae): 179. 1967.

Tree. Miombo woodland. 1300–1800 m.

Distribution: MP—*P. Lovett 59*

Brachystegia longifolia Benth.; Hoyle, F.T.E.A. (Leguminosae–Caesalpinioideae): 188. 1967.

Shrub or tree. Lakeshore thickets, woodland. 474–1400 m.

Distribution: MP, SL—*Mwasumbi 16583*

Brachystegia microphylla Harms; Hoyle, F.T.E.A. (Leguminosae–Caesalpinioideae): 174. 1967.

Tree. Miombo woodland. 1780–1790 m.

Distribution: SL—*Gereau 3871*

Brachystegia spiciformis Benth.; Hoyle, F.T.E.A. (Leguminosae–Caesalpinioideae): 167. 1967.

Tree. Miombo woodland, old cultivation. 1250–1790 m.

Distribution: MP, SL—*Kayombo 2370*

Brachystegia utilis Hutch. & Burtt Davy; Hoyle, F.T.E.A. (Leguminosae–Caesalpinioideae): 172. 1967.

Tree. Miombo woodland. 1300–1400 m.

Distribution: MP—*Kayombo 1042*

Brachystegia wangermeeana De Wild.; Hoyle, F.T.E.A. (Leguminosae–Caesalpinioideae): 192. 1967.

Tree. Miombo woodland. 1780 m.

Distribution: SL—*Gereau 5102*

Brachystegia sp.

Tree. Miombo woodland. 1780 m.

Distribution: SL—*Gereau 5099*

Caesalpinia decapetala (Roth) Alston; Brenan, F.T.E.A. (Leguminosae–Caesalpinioideae): 36. 1967.

Liana. Riverine forest, montane forest. 1400–1875 m.

Distribution: MP, RM—*J. Lovett 4690*

Cassia abbreviata Oliv. subsp. *abbreviata*; Brenan, F.T.E.A. (Leguminosae–Caesalpinioideae): 59. 1967.

Shrub. Miombo woodland. 980 m.

Distribution: SL—*Gereau 5159*

Cassia thyrsoidea Brenan; Brenan, F.T.E.A. (Leguminosae–Caesalpinioideae): 62. 1967.

Tree. Riverine forest edges. 1500 m.

Distribution: SL—*Gereau 3962*

Chamaecrista comosa E. Mey. var. *capricornia* (Steyaert) Lock; Lock, Fl. Zamb. 3(2): 143. 2007.

Terrestrial herb. Wooded grassland. 1860 m.

Distribution: SL—*Gereau 3930*

Chamaecrista mimosoides (L.) Greene; Lock, Fl. Zamb. 3(2): 136. 2007.

Terrestrial herb. Woodland, grassland, abandoned cultivation, secondary shrubland. 1250–1760 m.

Distribution: MP, SL—*Gereau 4379*

Chamaecrista parva (Steyaert) Lock; Lock, Fl. Zamb. 3(2): 147. 2007.

Terrestrial herb. Montane grassland. 2400–2640 m.

Distribution: KP—*Suleiman 53*

Chamaecrista stricta E. Mey.; Lock, Fl. Zamb. 3(2): 138. 2007.

Terrestrial herb. Riverine forest, miombo woodland, montane grassland. 1250–1940 m.

Distribution: MP, SL—*Gereau 4775*

Chamaecrista wittei (Ghesq.) Lock; Lock, Fl. Zamb. 3(2): 141. 2007.

Terrestrial herb. Montane brushland. 1450–2040 m.

Distribution: RM, SL—*Gereau 4750*

Julbernardia globiflora (Benth.) Troupin; Brenan, F.T.E.A. (Leguminosae–Caesalpinioideae): 147. 1967.

Tree. Miombo woodland. 1180–1400 m.

Distribution: MP, NL—*Kayombo 2411*

Pterolobium stellatum (Forssk.) Brenan; Brenan, F.T.E.A. (Leguminosae–Caesalpinioideae): 42. 1967.

Liana. Thicket woodland, riverine forest. 1275–1600 m.

Distribution: MP—*Kayombo 2479*

Senna occidentalis (L.) Link; Lock, Fl. Zamb. 3(2): 152. 2007.

Shrub. Riverine forest. 1200 m.

Distribution: MP—*J. Lovett 4405*

Senna septemtrionalis (Viv.) H. S. Irwin & Barneby; Lock, Fl. Zamb. 3(2): 158. 2007.

Shrub. Eroded hillsides, montane forest. 1300–1810 m.

Distribution: MP, RM, SL—*Gereau 4075*

Senna singueana (Delile) Lock; Lock, Fl. Zamb. 3(2): 160. 2007.

Tree. Woodland. 1200–1400 m.

Distribution: MP—*Kayombo 2405*

Tamarindus indica L.; Brenan, F.T.E.A. (Leguminosae–Caesalpinioideae): 153. 1967.

Tree. Lakeshore forest. 480 m.

Distribution: SL—*Gereau 3655*

CAMPANULACEAE

5 genera and 16 species

Canarina eminii Asch. ex Schweinf.; Thulin, F.T.E.A. (Campanulaceae): 4. 1976.

Epiphyte or vine. Montane forest. 1700–2360 m.

Distribution: MR, NL, RM—*Abeid 2774*

Cyphia brachyandra Thulin; Thulin, F.T.E.A. (Lobeliaceae): 52. 1984.

Herbaceous climber. Riverine forest. 1500 m.

Distribution: SL—*Gereau 4419*

Cyphia lasiandra Diels; Thulin, F.T.E.A. (Lobeliaceae): 55. 1984.

Herbaceous climber. Montane grassland. 2080–2400 m.

Distribution: KP, SL—*Suleiman 113*

Cyphia rupestris E. Wimm.; Thulin, F.T.E.A. (Lobeliaceae): 53. 1984.

Terrestrial herb. Montane forest edges, montane grassland. 2320–2740 m.

Distribution: KP, MP, SL—*Gereau 3972*

Cyphia ubenensis Engl.; Thulin, F.T.E.A. (Lobeliaceae): 52. 1984.

Terrestrial herb. Montane grassland.

Distribution: KP—*Leedal 5397*

Lobelia giberroa Hemsl.; Thulin, F.T.E.A. (Lobeliaceae): 9. 1984.

Giant rosette herb. Riverine forest, woodland, montane forest edges. 1810–2640 m.

Distribution: KP, MP, MR, NL, RM, SL—*Gereau 4651*

Lobelia mildbraedii Engl.; Thulin, F.T.E.A. (Lobeliaceae): 14. 1984.

Giant rosette herb. Swamps and stream banks in montane grassland. 2500–2560 m.

Distribution: KP—*Taylor 8524*

Lobelia trullifolia Hemsl. subsp. *trullifolia*; Thulin, F.T.E.A. (Lobeliaceae): 21. 1984.

Terrestrial herb. Marshes in montane grassland. 2550 m.

Distribution: KP—*Abeid 3116*

Monopsis stellarioides (C. Presl) Urb. subsp. *schimperiana* (Urb.) Thulin; Thulin, F.T.E.A. (Lobeliaceae): 47. 1984.

Terrestrial herb. Riverine forest. 1850 m.

Distribution: SL—*Gereau 4043*

Wahlenbergia abyssinica (Hochst. ex A. Rich.) Thulin subsp. *abyssinica*; Thulin, F.T.E.A. (Campanulaceae): 25. 1976.

Terrestrial herb. Montane grassland. 2400–2610 m.

Distribution: KP—*Suleiman 33*

Wahlenberbia capillacea (L. f.) A. DC. subsp. *tenuior* (Engl.) Thulin; Thulin, F.T.E.A. (Campanulaceae): 15. 1976.

Terrestrial herb. Montane grassland, lava flows. 2400–2740 m.

Distribution: KP, RM—*Suleiman 34*

Wahlenbergia capitata (Baker) Thulin; Thulin, F.T.E.A. (Campanulaceae): 19. 1976.

Terrestrial herb. Woodland. 1475 m.

Distribution: MP—*Kayombo 2394*

Wahlenbergia denticulata (Burch.) A. DC.; Thulin, F.T.E.A. (Campanulaceae): 15. 1976.

Terrestrial herb. Montane grassland. 2500 m.

Distribution: KP—*J. Lovett 779*

Wahlenbergia huttonii (Sond.) Thulin; Thulin, F.T.E.A. (Campanulaceae): 17. 1976.

Terrestrial herb. Montane grassland. 2488–2740 m.

Distribution: KP, SL—*Suleiman 169*

Wahlenbergia krebsii Cham. subsp. *arguta* (Hook. f.) Thulin; Thulin, F.T.E.A. (Campanulaceae): 13. 1976.

Terrestrial herb. Rocky montane grassland. 2560 m.

Distribution: KP, RM—*Gereau 3184*

Wahlenbergia virgata Engl.; Thulin, F.T.E.A. (Campanulaceae): 9. 1976.

Terrestrial herb. Woodland, wet grassland. 1300–2000 m.

Distribution: KP, MP, NL—*J. Lovett 4701*

CANNABACEAE

1 genus and 1 species

Cannabis sativa L.; Verdcourt, F.T.E.A. (Cannabaceae): 1. 1975.

Terrestrial herb. Edges of cultivated fields, disturbed montane grassland. 2200–2780 m.

Distribution: KP, MR—*Abeid 2808*

CANNACEAE

1 genus and 1 species

Canna indica L.; Lock, F.T.E.A. (Musaceae [Cannaceae]): 7. 1993.

Terrestrial herb. Montane forest. 1790 m.

Distribution: RM—*WCS SHCP 34*

CAPPARACEAE

5 genera and 5 species

Boscia salicifolia Oliv.; Elffers et al., F.T.E.A. (Capparidaceae): 52. 1964.

Tree. Miombo woodland. 1200 m.

Distribution: MP—*J. Lovett 3807*

Capparis tomentosa Lam.; Elffers et al., F.T.E.A. (Capparidaceae): 62. 1964.

Liana. Riverine forest. 1290 m.

Distribution: MP—*Gereau 6113*

Cleome monophylla L.; Elffers et al., F.T.E.A. (Capparidaceae): 5. 1964.

Terrestrial herb. Woodland, roadsides. 1300–1600 m.

Distribution: MP, SL—*Gereau 3951*

Maerua triphylla A. Rich. var. *pubescens* (Klotzsch) DeWolf; Elffers et al., F.T.E.A. (Capparidaceae): 45. 1964.

Shrub. Lakeshore forest. 474 m.

Distribution: SL—*Gereau 3662*

Ritchiea albersii Gilg; Elffers et al., F.T.E.A. (Capparidaceae): 21. 1964.

Shrub or tree. Montane forest. 1790 m.

Distribution: RM—*WCS SHCP 72*

CARICACEAE

1 genus and 1 species

Cylicomorpha parviflora Urb.; Hemsley, F.T.E.A. (Caricaceae): 2. 1958.

Tree. Montane forest. 1670 m.

Distribution: NL—*Bracebridge 57*

CARYOPHYLLACEAE

3 genera and 5 species

Cerastium dregeanum Fenzl; Sonder, 1860: 130.

Terrestrial herb. Rocky mountain slopes. 2800 m.

Distribution: KP—*Goldblatt 8216*

Note: New to East Africa, from South Africa.

Cerastium indicum Wight & Arn.; Turrill, F.T.E.A. (Caryophyllaceae): 19. 1956.

Terrestrial herb. Montane forest edges. 2320 m.

Distribution: SL—*Gereau 3970*

Cerastium octandrum Hochst. ex A. Rich.; Turrill, F.T.E.A. (Caryophyllaceae): 21. 1956.

Terrestrial herb. Montane grassland. 2570–2600 m.

Distribution: KP—*Gereau 3112*

Dianthus zeyheri Sond. subsp. *zeyheri*; Hooper, 1959: 47.

Terrestrial herb. Montane grassland. 2120–2488 m.

Distribution: SL—*Gereau 4310*

Note: New to East Africa, from South Africa (Transvaal Drakensberg).

Silene burchellii Otth ex DC.; Turrill, F.T.E.A. (Caryophyllaceae): 33. 1956.

Terrestrial herb. Montane grassland. 2250–2640 m.

Distribution: KP, MP—*Suleiman 46*

CECROPIACEAE

1 genus and 1 species

Myrianthus holstii Engl.; Berg, F.T.E.A. (Moraceae [Cecropiaceae]): 88. 1989.

Tree. Riverine forest, montane forest. 1350–2040 m.

Distribution: MP, RM, SL—*Gereau 4721*

CELASTRACEAE

4 genera and 9 species

Catha edulis Forssk.; Robson & Blakelock, F.T.E.A. (Celastraceae): 25. 1994.

Tree. Riverine forest, scree slopes, montane forest. 1500–2230 m.

Distribution: MR, RM, SL—*Mwasumbi 16176*

Gymnosporia buchananii Loes.; Jordaan & van Wyk, 2006: 521.

Tree. Riverine forest. 840 m.

Distribution: SL—*Kibure 1459*

Gymnosporia harveyana Loes. subsp. *stolzii* (N. Robson) Jordaan; Jordaan & van Wyk, 2004: 844.

Shrub, sometimes scandent. Montane forest. 2040–2100 m.

Distribution: RM, SL—*Gereau 4715*

Gymnosporia heterophylla (Eckl. & Zeyh.) Loes.; Jordaan & van Wyk, 2006: 520.

Shrub or tree. Montane forest, bamboo thicket, woodland. 1810–2380 m.

Distribution: KP, MP, NL, RM, SL—*Gereau 4128*

Gymnosporia livingstonei Jordaan; Jordaan & van Wyk, 2004: 851.

Subshrub. Wooded grassland. 1860 m.

Distribution: SL—*Gereau 3925* (type: K, MO)

Gymnosporia senegalensis (Lam.) Loes.; Jordaan & van Wyk, 2006: 519.

Tree or shrub. Woodland, montane forest edges. 1200–1780 m.

Distribution: MP, RM, SL—*Kayombo 2436*

Maytenus acuminata (L. f.) Loes.; Robson & Blakelock, F.T.E.A. (Celastraceae): 19. 1994.

Tree, shrub. Montane forest, forest edges. 1960–2260 m.

Distribution: MR, RM, SL—*Mwasumbi 16348*

Maytenus undata (Thunb.) Blakelock; Robson & Blakelock, F.T.E.A. (Celastraceae): 18. 1994.

Tree or shrub, sometimes scandent. Montane forest, bamboo forest, forest edges, riverine forest. 1300–2700 m.

Distribution: MP, NL, RM—*Kayombo 2476*

Mystroxylon aethiopicum (Thunb.) Loes.; Robson & Blakelock, F.T.E.A. (Celastraceae): 21. 1994.

Tree, shrub. Riverine forest, woodland, montane forest. 980–2320 m.

Distribution: MP, NL, RM, SL—*Suleiman 137*

CHENOPODIACEAE

1 genus and 3 species

Chenopodium ambrosioides L.; Brenan, F.T.E.A. (Chenopodiaceae): 10. 1954.

Terrestrial herb. Riverine forest. 1350 m.

Distribution: MP—*J. Lovett 4755*

Chenopodium procerum Hochst. ex Moq.; Brenan, F.T.E.A. (Chenopodiaceae): 11. 1954.

Terrestrial herb. Cultivated fields. 2280 m.

Distribution: MR—*Mwasumbi 16163*

Chenopodium schraderianum Schult.; Brenan, F.T.E.A. (Chenopodiaceae): 12. 1954.

Terrestrial herb. Cultivated fields. 1700 m.

Distribution: MP—*J. Lovett 3494*

CHRYSOBALANACEAE

2 genera and 3 species

Hirtella zanzibarica Oliv. subsp. *zanzibarica*; Prance, 2002: 994.

Tree. Riverine forest. 500–980 m.

Distribution: SL—*Kibure 1488*

Parinari curatellifolia Planch. ex Benth.; Prance & White, 1988: 112.

Tree. *Protea* woodland, miombo woodland, coffee plantations. 1300–2000 m.

Distribution: MP, SL—*Gereau 3853*

Parinari excelsa Sabine; Prance & White, 1988: 112.

Tree. Riverine forest, montane forest, bamboo thicket. 1620–2100 m.

Distribution: MP, NL, RM, SL—*Gereau 3928*

CLUSIACEAE

4 genera and 9 species

Garcinia buchananii Baker; Bamps et al., F.T.E.A. (Guttiferae): 24. 1978.

Tree. Riverine forest, woodland. 474–1400 m.

Distribution: MP, SL—*Mwasumbi 16591*

Garcinia kingaensis Engl.; Bamps et al., F.T.E.A. (Guttiferae): 22. 1978.

Shrub or tree. Riverine forest, montane forest. 980–2000 m.

Distribution: MP, RM, SL—*Gereau 5152*

Garcinia volkensii Engl.; Bamps et al., F.T.E.A. (Guttiferae): 23. 1978.

Tree. Montane forest. 1920 m.

Distribution: NL—*Bracebridge 49*

Harungana madagascariensis Lam. ex Poir.; Bamps et al., F.T.E.A. (Guttiferae): 32. 1978.

Shrub or tree. Montane forest. 2040 m.

Distribution: RM—*WCS SHCP 157*

Hypericum conjungens N. Robson; Bamps et al., F.T.E.A. (Guttiferae): 30. 1978.

Shrub. Montane forest. 2130 m.

Distribution: MR—*Mwasumbi 16191A*

Hypericum peplidifolium A. Rich.; Bamps et al., F.T.E.A. (Guttiferae): 31. 1978.

Terrestrial herb. Montane grassland, riverine forest. 1300–2590 m.

Distribution: KP, MP, MR, RM, SL—*Gereau 4430*

Hypericum quartinianum A. Rich.; Bamps et al., F.T.E.A. (Guttiferae): 29. 1978.

Shrub. Riverine forest. 1650 m.

Distribution: MP—*J. Lovett 4922*

Hypericum revolutum Vahl subsp. *revolutum*; Bamps et al., F.T.E.A. (Guttiferae): 28. 1978.

Shrub. Montane forest, rocky mountain slopes, stream banks in montane grassland. 2340–2800 m.

Distribution: KP, RM—*Abeid 3003*

Psorospermum febrifugum Spach var. *febrifugum*; Milne-Redhead, F.T.E.A. (Hypericaceae): 18. 1953.

Shrub. Miombo woodland, lakeside forest and thickets. 474–1450 m.

Distribution: MP, SL—*Gereau 5153*

Psorospermum febrifugum Spach var. *ferrugineum* (Hook. f.) Keay & Milne-Redh.; Milne-Redhead, F.T.E.A. (Hypericaceae): 18. 1953.

Shrub. Miombo woodland, grassland. 1470–1950 m.

Distribution: MP, MR, SL—*Gereau 3855*

COLCHICACEAE

2 genera and 2 species

Gloriosa superba L. var. *superba*; Hoenselaar, F.T.E.A. (Colchicaceae): 6. 2005.

Terrestrial herb or herbaceous climber. Riverine forest, montane forest, woodland, edges of cultivation. 1200–2140 m.

Distribution: MP, MR, NL, SL—*Suleiman 120*

Wurmbea tenuis (Hook. f.) Baker subsp. *goetzei* (Engl.) B. Nord.; Hoenselaar, F.T.E.A. (Colchicaceae): 14. 2005.

Terrestrial herb. Montane grassland, open woodland. 1720–2800 m.

Distribution: KP, MP—*Gereau 3497*

COMBRETACEAE

2 genera and 6 species

Combretum molle R. Br. ex G. Don; Wickens, F.T.E.A. (Combretaceae): 33. 1973.

Tree. Woodland, riverine forest. 1200–1680 m.

Distribution: MP, SL—*Kayombo 2406*

Combretum pentagonum M. A. Lawson; Wickens, F.T.E.A. (Combretaceae): 57. 1973.

Liana. Riverine forest, lakeshore shrubland. 474–1160 m.

Distribution: SL—*Gereau 3705*

Combretum psidioides Welw. subsp. *psidioides*; Wickens, F.T.E.A. (Combretaceae): 37. 1973

Tree. Woodland. 1300–1400 m.

Distribution: MP—*Kayombo 977*

Combretum zeyheri Sond.; Wickens, F.T.E.A. (Combretaceae): 31. 1973.

Tree or shrub. Grassland. 1300–1400 m.

Distribution: MP—*Kayombo 986*

Terminalia kaiseriana F. Hoffm.; Wickens, F.T.E.A. (Combretaceae): 84. 1973.

Tree. Open woodland. 1300–1400 m.

Distribution: MP—*Kayombo 1002*

Terminalia sambesiaca Engl. & Diels; Wickens, F.T.E.A. (Combretaceae): 91. 1973.

Tree. Lakeshore forest. 480 m.

Distribution: SL—*Gereau 3652*

COMMELINACEAE

5 genera and 16 species

Aneilema aequinoctiale (P. Beauv.) G. Don; Ensermu & Faden, 1997: 351.

Terrestrial herb. Riverine forest. 1650 m.

Distribution: MP—*J. Lovett 4921*

Aneilema dispermum Brenan; Brenan, F.W.T.A. 3: 31. 1968.

Terrestrial herb. Montane forest. 2000–2150 m.

Distribution: MR, RM—*Mwasumbi 16231*

Aneilema hirtum A. Rich.; Ensermu & Faden, 1997: 360.

Terrestrial herb. Woodland, riverine forest. 1250–1800 m.

Distribution: MP—*P. Lovett 222*

Aneilema pedunculosum C. B. Clarke; Faden, 1991: 147.

Terrestrial herb. Riverine forest. 1500 m.

Distribution: SL—*Gereau 4403*

Commelina africana L. subsp. *africana*; Faden, F.T.E.A. (Commelinaceae) [in press].

Terrestrial herb. Montane grassland, cultivated fields. 1825–2800 m.

Distribution: KP, MP, MR, NL, RM, SL—*Gereau 3586*

Commelina benghalensis L.; Ensermu & Faden, 1997: 373.

Terrestrial herb. Riverine forest, miombo woodland. 1210–1950 m.

Distribution: MP, SL—*Gereau 4290*

Commelina diffusa Burm. f. subsp. *diffusa*; Brenan, F.W.T.A. 3: 47. 1968.

Terrestrial herb. Montane forest. 1790 m.

Distribution: NL—*Gereau 4165*

Commelina hockii De Wild.; Faden, 2001: 403.

Terrestrial herb. Woodland, abandoned cultivation. 1250–1450 m.

Distribution: MP—*J. Lovett 3724*

Commelina kituloensis Faden; Faden, 2001: 400.

Terrestrial herb. Montane grassland, rocky ridges, woodland. 1600–2800 m.

Distribution: KP, MP, MR, NL, RM—*Faden 96/438* (type: BR, EA, K, MO, NHT, PRE, US), *Gereau 3181*

Commelina neurophylla C. B. Clarke; Clarke, F.T.A. 8: 53. 1901.

Terrestrial herb. Miombo woodland, montane grassland. 2000–2500 m.

Distribution: KP, MP—*P. Lovett 273*

Commelina subulata Roth; Ensermu & Faden, 1997: 364.

Terrestrial herb. Stream banks. 1825 m.

Distribution: MR—*Cribb 11389*

Commelina triangulispatha Mildbr.; Faden, 1994: 305.

Terrestrial herb. Roadside banks.

Distribution: KP—*Hepper 5318*

Cyanotis barbata D. Don; Ensermu & Faden, 1997: 341.

Terrestrial herb. Montane grassland. 2280–2600 m.

Distribution: KP, MP, MR—*Mwasumbi 16167*

Cyanotis longifolia Benth.; Ensermu & Faden, 1997: 340.

Terrestrial herb. Woodland. 1200–1600 m.

Distribution: MP—*J. Lovett 3621*

Floscopa glomerata (Willd. ex Schult. & Schult. f.) Hassk. subsp. *glomerata*; Brenan, F.W.T.A. 3: 28. 1968.

Terrestrial herb. Seasonally waterlogged grassland, woodland. 1250–1400 m.

Distribution: MP—*Kayombo 970*

Murdannia simplex (Vahl) Brenan; Ensermu & Faden, 1997: 346.

Terrestrial herb. Edges of agricultural fields. 1300 m.

Distribution: MP—*J. Lovett 3554*

CONNARACEAE

1 genus and 3 species

Rourea coccinea (Schumach. & Thonn.) Benth. subsp. *boiviniana* (Baill.) Jongkind; Jongkind, 1989: 332.

Scandent shrub. Disturbed montane forest.

Distribution: RM—*Ndangalasi 1026*

Rourea orientalis Baill.; Jongkind, 1989: 348.

Shrub. Miombo woodland, lakeshore shrubland. 474–980 m.

Distribution: SL—*Gereau 5154*

Rourea thomsonii (Baker) Jongkind; Jongkind, 1989: 359.

Liana. Montane forest. 1740–1780 m.

Distribution: RM—*Kibure 1372*

CONVOLVULACEAE

4 genera and 13 species

Astripomoea malvacea (Klotzsch) A. Meeuse var. *volkensii* (Dammer) Verdc.; Verdcourt, F.T.E.A. (Convolvulaceae): 76. 1963.

Trailing terrestrial herb. Edges of cultivation. 1350 m.

Distribution: MP—*J. Lovett 3533*

Dichondra repens J. R. Forst. & G. Forst.; Verdcourt, F.T.E.A. (Convolvulaceae): 12. 1963.

Creeping terrestrial herb. Montane forest.

Distribution: RM—*WCS SHCP 249*

Evolvulus alsinoides (L.) L.; Verdcourt, F.T.E.A. (Convolvulaceae): 18. 1963.

Terrestrial herb. Grassland. 530 m.

Distribution: SL—*Gilli 433*

Ipomoea alpina Rendle; Verdcourt, F.T.E.A. (Convolvulaceae): 104. 1963.

Terrestrial herb. Miombo woodland. 1530 m.

Distribution: SL—*Gereau 4365*

Ipomoea barteri Baker; Verdcourt, F.T.E.A. (Convolvulaceae): 119. 1963.

Herbaceous climber. Abandoned cultivation. 1250 m.

Distribution: MP—*J. Lovett 4072*

Ipomoea crassipes Hook. var. *crassipes*; Verdcourt, F.T.E.A. (Convolvulaceae): 98. 1963.

Trailing terrestrial herb. Woodland. 1200 m.

Distribution: MP—*J. Lovett 3878*

Ipomoea involucrata P. Beauv. var. *involucrata*; Verdcourt, F.T.E.A. (Convolvulaceae): 104. 1963.

Trailing terrestrial herb. Montane grassland, montane forest. 1940–2530 m.

Distribution: KP, SL—*Gereau 4774*

Ipomoea marginata (Desr.) Verdc.; Verdcourt, 1987: 658.

Herbaceous climber. Wet grassland. 1250 m.

Distribution: MP—*J. Lovett 4575*

Ipomoea pes-caprae (L.) R. Br. subsp. *brasiliensis* (L.) Ooststr.; Verdcourt, F.T.E.A. (Convolvulaceae): 121. 1963.

Creeping terrestrial herb. Sandy lakeshore. 474 m.

Distribution: SL—*Mwasumbi 16561*

Ipomoea prismatosyphon Welw. var. *prismatosyphon*; Verdcourt, F.T.E.A. (Convolvulaceae): 142. 1963.

Terrestrial herb. Woodland. 1250 m.

Distribution: MP—*J. Lovett 4615*

Ipomoea tenuirostris Choisy subsp. *tenuirostris*; Verdcourt, F.T.E.A. (Convolvulaceae): 101. 1963.

Straggling terrestrial herb. Woodland. 1250 m.

Distribution: MP, RM—*J. Lovett 4410*

Ipomoea welwitschii Vatke ex Hallier f.; Verdcourt, F.T.E.A. (Convolvulaceae): 119. 1963.

Trailing terrestrial herb. Woodland. 1250 m.

Distribution: MP—*J. Lovett 3651*

Ipomoea wightii (Wall.) Choisy var. *wightii*; Verdcourt, F.T.E.A. (Convolvulaceae): 110. 1963.

Herbaceous climber. Woodland. 1400–1500 m.

Distribution: MP—*J. Lovett 3414*

CORNACEAE

1 genus and 1 species

Cornus volkensii Harms; Lovett et al., 2006: 62.

Tree. Montane forest. 1760–2230 m.

Distribution: MR, RM, SL—*Gereau 4608*

COSTACEAE

1 genus and 1 species

Costus macranthus K. Schum.; Lock, F.T.E.A. (Zingiberaceae): 6. 1985.

Terrestrial herb. Woodland. 1250–1400 m.

Distribution: MP—*J. Lovett 3572*

CRASSULACEAE

2 genera and 5 species

Crassula cf. *alticola* R. Fern.; Fernandes, Fl. Zamb. 7(1): 15. 1983.

Terrestrial herb. Montane grassland. 2488 m.

Distribution: SL—*Suleiman 171*

Crassula schimperi Fisch. & C. A. Mey. subsp. *phyturus* (Mildbr.) R. Fern.; Wickens, F.T.E.A. (Crassulaceae): 8. 1987.

Terrestrial herb. Thin soil among boulders. 2600 m.

Distribution: MP—*Gereau 3504*

Crassula similis Baker f. vel sp. aff.; Fernandes, Fl. Zamb. 7(1): 29. 1983.

Terrestrial herb. Woodland, montane grassland. 1200–2488 m.

Distribution: MP, SL—*Suleiman 170*

Crassula vaginata Eckl. & Zeyh.; Wickens, F.T.E.A. (Crassulaceae): 13. 1987.

Terrestrial herb. Montane grassland, secondary woodland. 2000–2500 m.

Distribution: MP—*P. Lovett 502*

Crassula vaginata Eckl. & Zeyh. × *alba* Forssk.; Wickens, F.T.E.A. (Crassulaceae): 15. 1987.

Terrestrial herb. Montane grassland, roadsides. 2400–2730 m.

Distribution: KP, NL, RM—*Gereau 4099*

Kalanchoe crenata (Andrews) Haw. subsp.

nyassensis R. Fern.; Wickens, F.T.E.A. (Crassulaceae): 45. 1987.

Terrestrial herb. Montane forest. 2500 m.

Distribution: MP—*P. Lovett 485*

CUCURBITACEAE

11 genera and 15 species

Coccinia adoensis (Hochst. ex A. Rich.) Cogn.; Jeffrey, F.T.E.A. (Cucurbitaceae): 65. 1967.

Herbaceous climber. Montane forest remnants, riverine forest, lakeside forest, woodland, abandoned cultivation. 474–2160 m.

Distribution: MP, SL—*Gereau 3535*

Coccinia grandiflora Cogn.; Jeffrey, F.T.E.A. (Cucurbitaceae): 64. 1967.

Herbaceous climber. Abandoned cultivation. 1250 m.

Distribution: MP—*J. Lovett 3701*

Coccinia senensis (Klotzsch) Cogn.; Jeffrey, F.T.E.A. (Cucurbitaceae): 66. 1967.

Herbaceous climber. Riverine forest. 580 m.

Distribution: SL—*Gereau 3817*

Cucumella aetheocarpa C. Jeffrey vel sp. aff.; Jeffrey, F.T.E.A. (Cucurbitaceae): 114. 1967.

Herbaceous climber. Woodland. 1200 m.

Distribution: MP—*J. Lovett 3977*

Lagenaria abyssinica (Hook. f.) C. Jeffrey; Jeffrey, F.T.E.A. (Cucurbitaceae): 50. 1967.

Herbaceous climber. Woodland. 1300–1400 m.

Distribution: MP—*Kayombo 1003*

Momordica foetida Schumach.; Jeffrey, F.T.E.A. (Cucurbitaceae): 29. 1967.

Herbaceous climber. Woodland, montane forest edges. 1300–2250 m.

Distribution: MP, RM—*J. Lovett 3449*

Mukia maderaspatana (L.) M. Roem.; Jeffrey, F.T.E.A. (Cucurbitaceae): 115. 1967.

Herbaceous climber. Riverine forest, woodland, grassland. 1250–1400 m.

Distribution: MP—*Kayombo 996*

Oreosyce africana Hook. f.; Jeffrey, F.T.E.A. (Cucurbitaceae): 110. 1967.

Herbaceous climber. Abandoned cultivation, riverine forest, montane bushland, bamboo thicket. 1250–2300 m.

Distribution: MP, MR, NL, RM, SL—*Mwasumbi 16550*

Peponium vogelii (Hook. f.) Engl.; Jeffrey, F.T.E.A. (Cucurbitaceae): 81. 1967.

Herbaceous climber. Montane forest edges, riverine forest. 1500–2260 m.

Distribution: MP, SL—*Gereau 3531*

Raphidiocystis jeffreyana R. Fern. & A. Fern.; Jeffrey, F.T.E.A. (Cucurbitaceae): 55. 1967.

Herbaceous climber. Montane forest. 1710–1750 m.

Distribution: RM—*Abeid 2777*

Note: New to Tanzania, from Uganda, Democratic Republic of Congo, Cameroon, Gabon, and Angola.

Telfairia pedata (Sm. ex Sims) Hook.; Jeffrey, F.T.E.A. (Cucurbitaceae): 15. 1967.

Herbaceous climber. 1750 m.

Distribution: RM—*WCS SHCP 3*

Trochomeria macrocarpa (Sond.) Hook. f.; Jeffrey, F.T.E.A. (Cucurbitaceae): 87. 1967.

Straggling herb. Woodland, riverine forest. 1400–1500 m.

Distribution: MP—*J. Lovett 3294A*

Zehneria minutiflora (Cogn.) C. Jeffrey; Jeffrey, F.T.E.A. (Cucurbitaceae): 126. 1967.

Herbaceous climber. Montane forest. 1790 m.

Distribution: NL, RM—*Gereau 4159*

Zehneria cf. *oligosperma* C. Jeffrey; Jeffrey, F.T.E.A. (Cucurbitaceae): 124. 1967.

Herbaceous climber. Montane forest. 2000 m.

Distribution: MP—*P. Lovett 442*

Zehneria scabra (L. f.) Sond.; Jeffrey, F.T.E.A. (Cucurbitaceae): 122. 1967.

Herbaceous climber. Montane forest, grassland, coffee plantations. 1340–2400 m.

Distribution: MP, MR, RM—*Mwasumbi 16244*

CUSCUTACEAE

1 genus and 1 species

Cuscuta kilimanjari Oliv. var. *kilimanjari*; Verdcourt, F.T.E.A. (Convolvulaceae): 6. 1963.

Twining parasitic herb. Riverine forest. 1275–1400 m.

Distribution: MP—*Kayombo 2364*

CYPERACEAE

15 genera and 61 species

Ascolepis capensis (Kunth) Ridl.; Beentje, F.T.E.A. (Cyperaceae): 273. 2010.

Terrestrial herb. Montane grassland. 2560–2590 m.

Distribution: KP—*Gereau 2851*

Ascolepis hemisphaerica Peter ex Goetgh.; Beentje, F.T.E.A. (Cyperaceae): 273. 2010.

Terrestrial herb. Wooded grassland. 1270–1860 m.

Distribution: MP, SL—*Gereau 3488*

Bulbostylis atrosanguinea (Boeckeler) C. B. Clarke; Verdcourt, F.T.E.A. (Cyperaceae): 75. 2010.

Terrestrial herb. Rocky mountain slopes. 2700 m.

Distribution: KP—*Richards 18446*

Bulbostylis buchananii C. B. Clarke; Verdcourt, F.T.E.A. (Cyperaceae): 92. 2010.

Terrestrial herb. Woodland. 1200 m.

Distribution: MP—*J. Lovett 3999*

Bulbostylis densa (Wall.) Hand.–Mazz. subsp. *afromontana* (Lye) R. W. Haines; Verdcourt, F.T.E.A. (Cyperaceae): 100. 2010.

Terrestrial herb. Road banks in cultivation, montane grassland. 2200–2500 m.

Distribution: KP, MR—*Gereau 4533*

Bulbostylis filamentosa (Vahl) C. B. Clarke; Verdcourt, F.T.E.A. (Cyperaceae): 101. 2010.

Terrestrial herb. Miombo woodland. 1530 m.

Distribution: SL—*Gereau 4360*

Bulbostylis hispidula (Vahl) R. W. Haines subsp. *brachyphylla* (Cherm.) R. W. Haines; Verdcourt, F.T.E.A. (Cyperaceae): 82. 2010.

Terrestrial herb. Miombo woodland. 1790 m.

Distribution: SL—*Gereau 3896*

Bulbostylis hispidula (Vahl) R. W. Haines subsp. *filiformis* (C. B. Clarke) R. W. Haines; Verdcourt, F.T.E.A. (Cyperaceae): 83. 2010.

Terrestrial herb. Woodland. 1200 m.

Distribution: MP—*J. Lovett 3999B*

Bulbostylis hispidula (Vahl) R. W. Haines subsp. *hispidula*; Verdcourt, F.T.E.A. (Cyperaceae): 82. 2010.

Terrestrial herb. Abandoned cultivation. 1250 m.

Distribution: MP—*J. Lovett 4082*

Bulbostylis macra (Ridl.) C. B. Clarke; Verdcourt, F.T.E.A. (Cyperaceae): 72. 2010.

Terrestrial herb. Woodland. 1300–1600 m.

Distribution: MP—*J. Lovett 3615*

Bulbostylis pusilla (Hochst. ex A. Rich.) C. B. Clarke subsp. *congolensis* (De Wild.) R. W. Haines; Verdcourt, F.T.E.A. (Cyperaceae): 95. 2010.

Terrestrial herb. Woodland. 1200 m.

Distribution: MP—*J. Lovett 3999A*

Carex chlorosaccus C. B. Clarke; Verdcourt, F.T.E.A. (Cyperaceae): 432. 2010.

Terrestrial herb. Montane forest. 1790 m.

Distribution: NL—*Gereau 4155*

Carex cyrtosaccus C. B. Clarke; Verdcourt, F.T.E.A. (Cyperaceae): 444. 2010.

Terrestrial herb. Riverbanks in montane grassland. 2560 m.

Distribution: KP—*Gereau 3176*

Carex echinochloe Kunze subsp. *nyasensis* (C. B. Clarke) Lye; Verdcourt, F.T.E.A. (Cyperaceae): 431. 2010.

Terrestrial herb. Riverine forest. 1210 m.

Distribution: SL—*Gereau 4250*

Carex lycurus K. Schum. subsp. *lycurus*; Verdcourt, F.T.E.A. (Cyperaceae): 429. 2010.

Terrestrial herb. Riverine forest, riverbanks in montane grassland. 1500–2590 m.

Distribution: KP, MP—*Gereau 3155*

Carex petitiana A. Rich.; Verdcourt, F.T.E.A. (Cyperaceae): 439. 2010.

Terrestrial herb. Riverbanks and thickets in montane grassland. 2560–2650 m.

Distribution: KP—*Gereau 3193*

Carex spicatopaniculata C. B. Clarke; Verdcourt, F.T.E.A. (Cyperaceae): 431. 2010.

Terrestrial herb. Montane forest edges. 2230 m.

Distribution: SL—*Gereau 3573*

Carex steudneri Boeckeler; Verdcourt, F.T.E.A. (Cyperaceae): 435. 2010.

Terrestrial herb. Montane forest edges. 2320 m.

Distribution: KP, SL—*Gereau 3986*

Coleochloa setifera (Ridl.) Gilly var. *setifera*; Hoenselaar, F.T.E.A. (Cyperaceae): 377. 2010.

Terrestrial herb. Miombo woodland, montane forest/woodland transition. 1780–2140 m.

Distribution: SL—*Gereau 4006*

Cyperus ajax C. B. Clarke; Hoenselaar, F.T.E.A. (Cyperaceae): 202. 2010.

Terrestrial herb. Montane forest. 2200–2500 m.

Distribution: MP, RM—*Mwasumbi 16227*

Cyperus aff. *albopilosus* (C. B. Clarke) Kük.; Hoenselaar, F.T.E.A. (Cyperaceae): 155. 2010.

Terrestrial herb. Wooded grassland. 1860 m.

Distribution: SL—*Gereau 3926*

Cyperus amabilis Vahl; Hoenselaar, F.T.E.A. (Cyperaceae): 158. 2010.

Terrestrial herb. Abandoned cultivation. 1250 m.

Distribution: MP—*J. Lovett 4173*

Cyperus articulatus L.; Hoenselaar, F.T.E.A. (Cyperaceae): 208. 2010.

Terrestrial herb. 1970 m.

Distribution: RM—*WCS SHCP 52*

Cyperus cyperoides (L.) Kuntze; Beentje, F.T.E.A. (Cyperaceae): 223. 2010.

Terrestrial herb. Montane forest, grassland, riverine forest, woodland, roadsides. 474–2150 m.

Distribution: MP, NL, SL—*Suleiman 165*

Cyperus dichrostachyus Hochst. ex A. Rich.; Hoenselaar, F.T.E.A. (Cyperaceae): 198. 2010.

Terrestrial herb. Swampy depressions in montane grassland, riverine forest. 1500–2590 m.

Distribution: KP, NL—*Gereau 2847*

Cyperus distans L. f.; Beentje, F.T.E.A. (Cyperaceae): 249. 2010.

Terrestrial herb. Riverine forest, montane grassland, woodland/grassland transition, seasonal ponds, forest clearings, abandoned cultivation. 1160–2230 m.

Distribution: MP, NL, SL—*Gereau 4270*

Cyperus esculentus L.; Hoenselaar, F.T.E.A. (Cyperaceae): 227. 2010.

Terrestrial herb. Miombo woodland, riverine forest, edges of cultivation, grassland, roadsides. 474–1500 m.

Distribution: MP, SL—*Gereau 3957*

Cyperus exaltatus Retz. var. *dives* (Delile) C. B. Clarke; Beentje, F.T.E.A. (Cyperaceae): 245. 2010.

Terrestrial herb. Woodland, inundated grassland. 1270–1400 m.

Distribution: MP—*Gereau 3492*

Cyperus glaucophyllus Boeckeler; Beentje, F.T.E.A. (Cyperaceae): 198. 2010.

Terrestrial herb. Forest clearings, forest/woodland transition, scree slopes, riverine forest. 1850–2140 m.

Distribution: NL, SL—*Gereau 4001*

Cyperus involucratus Rottb.; Hoenselaar, F.T.E.A. (Cyperaceae): 187. 2010.

Terrestrial herb. Riverine forest. 980–1400 m.

Distribution: MP, NL—*Mwangoka 663*

Cyperus laxus Lam. subsp. *sylvestris* (Ridl.) Lye; Hoenselaar, F.T.E.A. (Cyperaceae): 201. 2010.

Terrestrial herb. Montane forests, riverbanks. 1200–2230 m.

Distribution: MP, SL—*Gereau 4428*

Cyperus luteus Boeckeler; Hoenselaar, F.T.E.A. (Cyperaceae): 171. 2010.

Terrestrial herb. Woodland, cultivated fields. 1350–2200 m.

Distribution: MP, MR—*Gereau 4525*

Cyperus mollipes (C. B. Clarke) K. Schum.; Hoenselaar & Beentje, F.T.E.A. (Cyperaceae): 180. 2010.

Terrestrial herb. Riverine forest, miombo woodland, montane grassland. 580–2230 m.

Distribution: SL—*Gereau 3596*

Cyperus niveus Retz. var. *leucocephalus* (Kunth) Fosberg; Hoenselaar, F.T.E.A. (Cyperaceae): 170. 2010.

Terrestrial herb. Woodland. 1200 m.

Distribution: MP—*J. Lovett 3986*

Cyperus nyererei Lye; Hoenselaar, F.T.E.A. (Cyperaceae): 176. 2010.

Terrestrial herb. Woodland, montane grassland, scree slopes. 1600–2740 m.

Distribution: KP, MP, SL—*Wingfield 529* (type: DSM, K), *Gereau 3507*

Cyperus pluribracteatus (Kük.) Govaerts; Hoenselaar, F.T.E.A. (Cyperaceae): 223. 2010.

Terrestrial herb. Woodland. 1200 m.

Distribution: MP—*J. Lovett 4048*

Cyperus rigidifolius Steud.; Hoenselaar, F.T.E.A. (Cyperaceae): 165. 2010.

Terrestrial herb. Montane grassland. 1810–2600 m.

Distribution: KP, MP, MR, SL—*Gereau 2852*

Cyperus rotundus L.; Beentje, F.T.E.A. (Cyperaceae): 211. 2010.

Terrestrial herb. Grassland, riverine forest. 1500–2230 m.

Distribution: MP, SL—*Gereau 3592*

Cyperus tenuiculmis Boeckeler var. *guineensis* (Nelmes) S. S. Hooper; Hoenselaar, F.T.E.A. (Cyperaceae): 243. 2010.

Terrestrial herb. Seasonal ponds in montane grassland. 1840 m.

Distribution: SL—*Gereau 3915*

Cyperus tenuiculmis Boeck. var. *tenuiculmis*; Hoenselaar, F.T.E.A. (Cyperaceae): 243. 2010.

Terrestrial herb. Riverine forest, montane grassland. 1160–2120 m.

Distribution: SL—*Gereau 4316*

Cyperus tomaiophyllus K. Schum.; Hoenselaar, F.T.E.A. (Cyperaceae): 239. 2010.

Terrestrial herb. Montane forest edges. 2360 m.

Distribution: MR—*Gereau 3783*

Ficinia filiformis (Lam.) Schrad.; Muasya, F.T.E.A. (Cyperaceae): 118. 2010.

Terrestrial herb. Montane grassland, rocky outcrops. 2610–2800 m.

Distribution: KP, RM—*Gereau 2822*

Fimbristylis complanata (Retz.) Link; Verdcourt, F.T.E.A. (Cyperaceae): 50. 2010

Terrestrial herb. Grassland, woodland. 1250–1270 m.

Distribution: MP—*Gereau 3483*

Fimbristylis dichotoma (L.) Vahl; Verdcourt, F.T.E.A. (Cyperaceae): 57. 2010.

Terrestrial herb. Miombo woodland, riverine forest. 474–1800 m.

Distribution: MP, NL, SL—*Gereau 3682*

Isolepis costata Hochst. ex A. Rich.; Muasya, F.T.E.A. (Cyperaceae): 123. 2010.

Terrestrial herb. Riverbanks in montane grassland. 2550 m.

Distribution: KP—*Gereau 3149*

Kyllinga brevifolia Rottb. var. *lurida* (Kük.) Beentje; Beentje, F.T.E.A. (Cyperaceae): 315. 2010.

Terrestrial herb. Montane grassland. 2230–2360 m.

Distribution: KP, MR, SL—*Gereau 3740*

Kyllinga comosipes (Mattf. & Kük.) Napper; Beentje, F.T.E.A. (Cyperaceae): 338. 2010.

Terrestrial herb. Thin soil among boulders. 2600 m.

Distribution: MP—*Gereau 3506*

Kyllinga crassipes Boeckeler; Beentje, F.T.E.A. (Cyperaceae): 342. 2010.

Terrestrial herb. Riverine forest. 1160–1500 m.

Distribution: SL—*Gereau 3941*

Kyllinga odorata Vahl var. *major* (C. B. Clarke) Chiov.; Beentje, F.T.E.A. (Cyperaceae): 340. 2010.

Terrestrial herb. Montane forest, abandoned cultivation. 1250–2110 m.

Distribution: MP, MR, NL—*Gereau 4563*

Kyllinga odorata Vahl var. *odorata*; Beentje, F.T.E.A. (Cyperaceae): 340. 2010.

Terrestrial herb. Montane forest. 1900 m.

Distribution: NL—*Gereau 4218*

Kyllinga tanzaniae Lye; Beentje, F.T.E.A. (Cyperaceae): 343. 2010.

Terrestrial herb. Grassland, miombo woodland. 1270–1400 m.

Distribution: MP—*Gereau 3485*

Kyllinga tenuifolia Steud.; Beentje, F.T.E.A. (Cyperaceae): 29. 2010.

Terrestrial herb. Riverine forest, miombo woodland. 1200 m.

Distribution: MP—*J. Lovett 3990*

Kyllingiella polyphylla (A. Rich.) Lye; Beentje, F.T.E.A. (Cyperaceae): 128. 2010.

Terrestrial herb. Grassland. 1270 m.

Distribution: MP—*Gereau 3489*

Lipocarpha comosa J. Raynal vel sp. aff.; Beentje, F.T.E.A. (Cyperaceae): 347. 2010.

Terrestrial herb. Woodland. 1250 m.

Distribution: MP—*J. Lovett 4332*

Pycreus capillifolius (A. Rich.) C. B. Clarke; Hoenselaar, F.T.E.A. (Cyperaceae): 284. 2010.

Terrestrial herb. Road banks in cultivation. 2200 m.

Distribution: MR—*Gereau 4534*

Pycreus elegantulus (Steud.) C. B. Clarke; Hoenselaar, F.T.E.A. (Cyperaceae): 303. 2010.

Terrestrial herb. Swampy depressions in montane grassland, riverine forest, abandoned cultivation. 1250–2590 m.

Distribution: KP, MP, SL—*Gereau 2848*

Pycreus flavescens (L.) P. Beauv. ex Rchb.; Hoenselaar, F.T.E.A. (Cyperaceae): 292. 2010.

Terrestrial herb. Riverine forest, wooded grassland. 1200–1500 m.

Distribution: MP, SL—*Gereau 3943*

Pycreus macrostachyos (Lam.) J. Raynal; Hoenselaar, F.T.E.A. (Cyperaceae): 288. 2010.

Terrestrial herb. Riverine forest, wooded grassland. 1250–1350 m.

Distribution: MP—*J. Lovett 4658*

Schoenoplectus corymbosus (Roth ex Roem. & Schult.) J. Raynal; Beentje, F.T.E.A. (Cyperaceae): 25. 2010.

Terrestrial herb. Seasonal ponds in montane grassland. 1840 m.

Distribution: SL—*Gereau 3918*

Schoenoxiphium ludwigii Hochst.; Verdcourt, F.T.E.A. (Cyperaceae): 420. 2010.

Terrestrial herb. Riverbanks in montane grassland. 2560 m.

Distribution: KP—*Gereau 3172*

Schoenoxiphium sparteum (Wahlenb.) C. B. Clarke; Verdcourt, F.T.E.A. (Cyperaceae): 418. 2010.

Terrestrial herb.

Distribution: MP—*Napper 1696*

Scleria bulbifera Hochst. ex A. Rich.; Verdcourt, F.T.E.A. (Cyperaceae): 386. 2010.

Terrestrial herb. Miombo woodland. 1200–1600 m.

Distribution: MP, SL—*Gereau 4307*

Scleria lagoensis Boeckeler; Verdcourt, F.T.E.A. (Cyperaceae): 407. 2010.

Terrestrial herb. Miombo woodland. 1260 m.

Distribution: SL—*Gereau 4306*

Scleria racemosa Poir.; Verdcourt, F.T.E.A. (Cyperaceae): 413. 2010.

Terrestrial herb. Riverine forest. 474–1500 m.

Distribution: NL—*Gereau 5169*

Scleria woodii C. B. Clarke var. *woodii*; Verdcourt, F.T.E.A. (Cyperaceae): 391. 2010.

Terrestrial herb. Grassland. 1270 m.

Distribution: MP—*Gereau 3484*

DIOSCOREACEAE

1 genus and 5 species

Dioscorea buchananii Benth.; Milne-Redhead, F.T.E.A. (Dioscoreaceae): 19. 1975.

Herbaceous climber. Woodland. 1300–1600 m.

Distribution: MP—*J. Lovett 4286*

Dioscorea cochleariapiculata De Wild.; Milne-Redhead, F.T.E.A. (Dioscoreaceae): 22. 1975.

Herbaceous climber. Woodland. 1250–1400 m.

Distribution: MP—*Kayombo 962*

Dioscorea dumetorum (Kunth) Pax; Milne-Redhead, F.T.E.A. (Dioscoreaceae): 21. 1975.

Herbaceous climber. Riverine forest, woodland. 1160–1200 m.

Distribution: MP, SL—*Gereau 3692*

Dioscorea quartiniana A. Rich.; Milne-Redhead, F.T.E.A. (Dioscoreaceae): 23. 1975.

Herbaceous climber. Riverine forest, montane forest. 1210–1850 m.

Distribution: NL, SL—*Gereau 4285*

Dioscorea schimperiana Hochst. ex Kunth; Milne-Redhead, F.T.E.A. (Dioscoreaceae): 14. 1975.

Herbaceous climber. Riverine forest, woodland. 1200–1500 m.

Distribution: MP, NL, SL—*Gereau 3948*

DIPSACACEAE

2 genera and 2 species

Cephalaria pungens Szabó; Napper, F.T.E.A. (Dipsacaceae): 4. 1968.

Terrestrial herb. Montane grassland. 2500–2700 m.

Distribution: KP, MP, NL—*Gereau 4098*

Scabiosa austroafricana Heine; Napper, F.T.E.A. (Dipsacaceae): 9. 1968.

Terrestrial herb. Montane grassland, woodland. 1720–2600 m.

Distribution: KP, MP—*Gereau 3088*

DIPTEROCARPACEAE

1 genus and 2 species

Monotes adenophyllus Gilg subsp. *adenophyllus*; Verdcourt, F.T.E.A. (Dipterocarpaceae): 7. 1989.

Tree. Woodland. 1200–1310 m.

Distribution: MP—*J. Lovett 4312*

Monotes africanus A. DC.; Verdcourt, F.T.E.A. (Dipterocarpaceae): 8. 1989.

Tree. Miombo woodland. 1530–1790 m.

Distribution: SL—*Gereau 3873*

DRACAENACEAE

2 genera and 5 species

Dracaena afromontana Mildbr.; Mwachala, F.T.E.A. (Dracaenaceae): 6. 2007.

Tree. Montane forest. 1985–2380 m.

Distribution: KP, RM, SL—*Abeid 2816*

Dracaena laxissima Engl.; Mwachala, F.T.E.A. (Dracaenaceae): 3. 2007.

Shrub, liana. Riverine forest, montane forest. 1810–2110 m.

Distribution: MP, MR, SL—*Gereau 4555*

Dracaena mannii Baker; Mwachala, F.T.E.A. (Dracaenaceae): 7. 2007.

Shrub. Lakeshore bushland, montane forest. 485–2000 m.

Distribution: RM, SL—*Mwasumbi 16613*

Dracaena steudneri Engl.; Mwachala, F.T.E.A. (Dracaenaceae): 8. 2007.

Tree or shrub. Woodland, riverine forest. 1300–2000 m.

Distribution: MP, NL, RM—*Gereau 6108*

Sansevieria kirkii Baker; Mbugua, F.T.E.A. (Dracaenaceae): 30. 2007.

Terrestrial herb or suffrutex. Riverine forest, woodland. 580–1250 m.

Distribution: MP, SL—*Gereau 3825*

DROSERACEAE

1 genus and 2 species

Drosera madagascariensis DC.; Laundon, F.T.E.A. (Droseraceae): 5. 1959.

Terrestrial herb. Stream bank in montane grassland. 2740 m.

Distribution: KP—*Abeid 3067*

Drosera pilosa Exell & J. R. Laundon; Laundon, F.T.E.A. (Droseraceae): 4. 1959.

Terrestrial herb. Montane grassland. 2488–2700 m.

Distribution: NL, RM, SL—*Gereau 4086*

EBENACEAE

2 genera and 6 species

Diospyros abyssinica (Hiern) F. White subsp. *abyssinica*; White & Verdcourt, F.T.E.A. (Ebenaceae): 22. 1996.

Tree. Montane forest. 1900 m.

Distribution: RM—*Bracebridge 45*

Diospyros mespiliformis Hochst. ex A. DC.; White & Verdcourt, F.T.E.A. (Ebenaceae): 40. 1996.

Tree. Valley forest. 1200 m.

Distribution: MP—*J. Lovett 4248*

Diospyros natalensis (Harv.) Brenan; White & Verdcourt, F.T.E.A. (Ebenaceae): 13. 1996.

Shrub. Riverine forest. 980 m.

Distribution: SL—*Gereau 5155*

Diospyros whyteana (Hiern) F. White; White & Verdcourt, F.T.E.A. (Ebenaceae): 28. 1996.

Tree or shrub. Montane forest. 1750–2380 m.

Distribution: KP, MR, NL, RM, SL—*Mwasumbi 16194*

Diospyros zombensis (B. L. Burtt) F. White; White & Verdcourt, F.T.E.A. (Ebenaceae): 27. 1996.

Tree or shrub. Riverine forest. 860–1210 m.

Distribution: SL—*Kibure 1450*

Euclea racemosa L. subsp. *schimperi* (A. DC.) F. White; White & Verdcourt, F.T.E.A. (Ebenaceae): 46. 1996.

Shrub. Miombo woodland, riverine forest. 474–1530 m.

Distribution: MP, NL, SL—*Gereau 4799*

ERICACEAE

2 genera and 6 species

Agarista salicifolia (Comm. ex Lam.) G. Don; Beentje, F.T.E.A. (Ericaceae): 2. 2006.

Tree. Miombo woodland, tree clumps in montane grassland. 1730–2800 m.

Distribution: KP, MR, RM, SL—*Gereau 5096*

Erica benguelensis (Welw. ex Engl.) E. G. H. Oliv. var. *albescens* (R. Ross) E. G. H. Oliv.; Beentje, F.T.E.A. (Ericaceae): 10. 2006.

Tree. Montane shrubland. 2100–2540 m.

Distribution: KP, MR, NL—*Abeid 2873*

Erica benguelensis (Welw. ex Engl.) E. G. H. Oliv. var. *benguelensis*; Beentje, F.T.E.A. (Ericaceae): 9. 2006.

Tree. Montane heathland and forest edges. 2340–2350 m.

Distribution: RM—*Abeid 2990*

Erica kingaensis Engl. subsp. *bequaertii* (De Wild.) R. Ross; Beentje, F.T.E.A. (Ericaceae): 19. 2006.

Shrub. Stream banks in montane grassland. 2550 m.

Distribution: KP—*Gereau 3157*

Note: New to Tanzania, from Rwenzori Mountains in Uganda and Democratic Republic of Congo.

Erica kingaensis Engl. subsp. *kingaensis*; Beentje, F.T.E.A. (Ericaceae): 18. 2006.

Shrub or subshrub. Grassland, heathland, secondary shrubland. 1760–2770 m.

Distribution: KP, MR, NL, RM—*Mwasumbi 16254*

Erica mannii (Hook. f.) Beentje subsp. *pallidiflora* (Engl.) E. G. H. Oliv.; Beentje, F.T.E.A. (Ericaceae): 17. 2006.

Shrub. Woodland. 1300–1790 m.

Distribution: MP, SL—*Kayombo 2481*

Erica mannii (Hook. f.) Beentje subsp. *usambarensis* (Alm. & T. C. E. Fr.) Beentje; Beentje, F.T.E.A. (Ericaceae): 17. 2006.

Shrub. Rocky mountaintop grassland. 2960 m.

Distribution: RM—*Abeid 3005*

Erica silvatica (Engl.) Beentje; Beentje, F.T.E.A. (Ericaceae): 19. 2006.

Subshrub. Heathland, montane grassland. 1940–2770 m.

Distribution: KP, MR, NL, RM, SL—*Mwasumbi 16261*

Erica whyteana Britten; Beentje, F.T.E.A. (Ericaceae): 11. 2006.

Subshrub. Stream banks and swamps in montane grassland. 2500–2830 m.

Distribution: KP—*Abeid 2856*

ERIOCAULACEAE

1 genus and 1 species

Eriocaulon schimperi Körn. ex Ruhland; Phillips, F.T.E.A. (Eriocaulaceae): 15. 1997.

Terrestrial herb. Ravines and stream banks in montane grassland. 2560–2600 m.

Distribution: KP, MP—*Gereau 3116*

ERIOSPERMACEAE

1 genus and 1 species

Eriospermum abyssinicum Baker; Whitehouse, F.T.E.A. (Eriospermaceae): 4. 1996.

Terrestrial herb. Miombo woodland, rocky mountain slopes. 1200–2800 m.

Distribution: KP, MP—*J. Lovett 3811*

ERYTHROXYLACEAE

1 genus and 1 species

Erythroxylum emarginatum Thonn.; Verdcourt, F.T.E.A. (Erythroxylaceae): 3. 1984.

Shrub. Riverine forest. 1320 m.

Distribution: MP—*Gereau 6125*

EUPHORBIACEAE

19 genera and 50 species

Acalypha fimbriata Schumach. & Thonn.; Radcliffe-Smith, F.T.E.A. (Euphorbiaceae 1): 198. 1987.

Terrestial herb. Roadsides. 474–490 m.

Distribution: SL—*Gereau 3792*

Acalypha ornata Hochst. ex A. Rich.; Radcliffe-Smith, F.T.E.A. (Euphorbiaceae 1): 190. 1987.

Shrub. Riverine forest, woodland, abandoned cultivation. 1160–1500 m.

Distribution: MP, SL—*Gereau 3700*

Acalypha petiolaris Hochst.; Prain & Hutchinson,

1913: 18.

Terrestrial herb. Woodland, cultivated fields, lakeshore bushland. 474–1600 m.

Distribution: MP, SL—*Mwasumbi 16577*

Acalypha polymorpha Müll. Arg.; Hutchinson, F.T.A. 6(1): 894. 1912.

Terrestrial herb. Grassland, woodland, abandoned cultivation. 1250–1940 m.

Distribution: MP, SL—*Gereau 4759*

Acalypha psilostachya Hochst. var. *glandulosa* Hutch.; Radcliffe-Smith, F.T.E.A. (Euphorbiaceae 1): 204. 1987.

Shrub, often scandent, or perennial herb. Montane forest, riverine forest, scree slopes, woodland. 1500–2400 m.

Distribution: MP, MR, SL—*Gereau 3637*

Acalypha psilostachya Hochst. var. *psilostachya*; Radcliffe-Smith, F.T.E.A. (Euphorbiaceae 1): 204. 1987.

Shrub, often scandent, or perennial herb. Montane forest, tree clumps in grassland, riverine forest, roadsides in cultivation, tree plantations. 1160–2460 m.

Distribution: KP, MP, MR, NL, RM, SL—*Mwasumbi 16226*

Antidesma membranaceum Müll. Arg.; Radcliffe-Smith, F.T.E.A. (Euphorbiaceae 2): 574. 1988.

Tree or shrub. Abandoned cultivation. 1250 m.

Distribution: MP—*J. Lovett 3748*

Antidesma venosum E. Mey. ex Tul.; Radcliffe-Smith, F.T.E.A. (Euphorbiaceae 2): 573. 1988.

Tree or shrub. Woodland, riverine forest, lakeshore forest. 474–1350 m.

Distribution: MP, SL—*Gereau 3665*

Bridelia brideliifolia (Pax) Fedde; Radcliffe-Smith, F.T.E.A. (Euphorbiaceae 1): 126. 1987.

Tree. Montane forest, woodland/forest transition. 1900–2110 m.

Distribution: MR, NL, SL—*Gereau 4743*

Bridelia cathartica G. Bertol.; Radcliffe-Smith, F.T.E.A. (Euphorbiaceae 1): 123. 1987.

Shrub. Miombo woodland, wooded grassland, riverine forest. 1200–1400 m.

Distribution: MP—*J. Lovett 4359*

Bridelia micrantha (Hochst.) Baill.; Radcliffe-Smith, F.T.E.A. (Euphorbiaceae 1): 127. 1987.

Tree. Montane forest, woodland, riverine forest, edges of cultivation. 1160–1920 m.

Distribution: MP, NL, RM, SL—*Kibure 1402*

Bridelia scleroneura Müll. Arg.; Radcliffe-Smith, F.T.E.A. (Euphorbiaceae 1): 122. 1987.

Shrub. Miombo woodland. 1200 m.

Distribution: MP—*Kayombo 2409*

Clutia abyssinica Jaub. & Spach var. *abyssinica*; Radcliffe-Smith, F.T.E.A. (Euphorbiaceae 1): 334. 1987.

Shrub. Montane forest, woodland, abandoned cultivation. 1250–2470 m.

Distribution: KP, MP, MR, NL, SL—*Gereau 4712*

Clutia abyssinica Jaub. & Spach var. *pedicellaris* (Pax) Pax; Radcliffe-Smith, F.T.E.A. (Euphorbiaceae 1): 336. 1987.

Shrub. Montane forest, forest edges, miombo woodland. 1400–2400 m.

Distribution: MP, MR, RM—*Mwasumbi 16181*

Clutia angustifolia Knauf; Radcliffe-Smith, F.T.E.A. (Euphorbiaceae 1): 339. 1987.

Shrub. Miombo woodland, grassland. 980–1940 m.

Distribution: NL, SL—*Gereau 4653*

Clutia paxii Knauf; Radcliffe-Smith, F.T.E.A. (Euphorbiaceae 1): 338. 1987.

Shrub. Montane forest, stream banks in montane grassland. 2400–2665 m.

Distribution: KP, MP—*Gereau 3126*

Clutia whytei Hutch.; Radcliffe-Smith, F.T.E.A. (Euphorbiaceae 1): 333. 1987.

Shrub. Wooded grassland. 1860 m.

Distribution: SL—*Gereau 3924*

Clutia sp.

Shrub. Montane forest edges, scree slopes. 1980–2120 m.

Distribution: SL—*Gereau 4330*

Clutia sp.

Shrub. Riverbanks in montane grassland. 2500–2660 m.

Distribution: KP—*Taylor 8520*

Croton macrostachyus Hochst. ex Delile; Radcliffe-Smith, F.T.E.A. (Euphorbiaceae 1): 149. 1987.

Tree. Woodland, riverine forest. 1300–1850 m.

Distribution: MP, RM, SL—*Gereau 4025*

Croton scheffleri Pax; Radcliffe-Smith, F.T.E.A. (Euphorbiaceae 1): 147. 1987.

Shrub. Riverine forest. 1200 m.

Distribution: MP—*J. Lovett 3836*

Drypetes gerrardii Hutch. var. *gerrardii*; Radcliffe-Smith, F.T.E.A. (Euphorbiaceae 1): 98. 1987.

Tree. Montane forest. 1900–2260 m.

Distribution: NL, RM—*Abeid 2778*

Drypetes gerrardii Hutch. var. *tomentosa* Radcl.-Sm.; Radcliffe-Smith, F.T.E.A. (Euphorbiaceae 1): 98. 1987.

Tree. Riverine forest. 1450 m.

Distribution: MP—*J. Lovett 4898*

Drypetes natalensis (Harv.) Hutch. var. *natalensis*; Radcliffe-Smith, F.T.E.A. (Euphorbiaceae 1): 92. 1987.

Tree. Montane forest. 1780–1955 m.

Distribution: NL, RM—*Kibure 1428*

Erythrococca trichogyne (Müll. Arg.) Prain; Radcliffe-Smith, F.T.E.A. (Euphorbiaceae 1): 274. 1987.

Shrub. Montane forest. 1780–2300 m.

Distribution: MR, RM, SL—*Mwasumbi 16539*

Euphorbia daviesii E. A. Bruce; Carter, F.T.E.A. (Euphorbiaceae 2): 437. 1988.

Terrestrial herb. Montane grassland. 1810–2800 m.

Distribution: KP, SL—*R. M. Davies E.17* (type: K), *Harder 1493*

Euphorbia grantii Oliv.; Carter, F.T.E.A. (Euphorbiaceae 2): 457. 1988.

Shrub. 1400 m.

Distribution: MP—*Michelmore 973*

Euphorbia heterophylla L.; Carter, F.T.E.A. (Euphorbiaceae 2): 431. 1988.

Terrestrial herb. Abandoned cultivation, lakeshore, riverine forest. 474–1500 m.

Distribution: MP, SL—*Gereau 4397*

Euphorbia hirta L.; Carter, F.T.E.A. (Euphorbiaceae 2): 415. 1988.

Terrestrial herb. Riverine forest, roadsides. 474–1350 m.

Distribution: MP, SL—*Gereau 3787*

Euphorbia ugandensis Pax; Carter, F.T.E.A. (Euphorbiaceae 2): 441. 1988.

Shrub. Montane forest. 2040–2640 m.

Distribution: MR, NL, RM, SL—*Gereau 4708*

Flueggea virosa (Roxb. ex Willd.) Royle; Radcliffe-Smith, F.T.E.A. (Euphorbiaceae 1): 68. 1987.

Shrub. Riverine forest, lakeshore forest, abandoned cultivation. 480–1500 m.

Distribution: MP, SL—*Gereau 3937*

Macaranga capensis (Baill.) Benth. ex Sim var. *capensis*; Lovett et al., 2006: 85.

Tree. Montane forest, riverine forest. 840–1760 m.

Distribution: RM, SL—*Kibure 1436*

Macaranga capensis (Baill.) Benth. ex Sim var. *kilimandscharica* (Pax) Friis & M. G. Gilbert; Lovett et al., 2006: 85.

Tree. Montane forest. 1900–2230 m.

Distribution: MR, NL, SL—*Gereau 4215*

Macaranga conglomerata Brenan; Radcliffe-Smith, F.T.E.A. (Euphorbiaceae 1): 246. 1987.

Tree. Miombo woodland. 620 m.

Distribution: SL—*Harder 1523*

Margaritaria discoidea (Baill.) G. L. Webster var. *nitida* (Pax) Radcl.-Sm.; Radcliffe-Smith, F.T.E.A. (Euphorbiaceae 1): 67. 1987.

Tree or shrub. Woodland, riverine forest. 580–1200 m.

Distribution: MP, SL—*Harder 1510*

Margaritaria discoidea (Baill.) G. L. Webster var. *triplosphaera* Radcl.-Sm.; Radcliffe-Smith, F.T.E.A. (Euphorbiaceae 1): 67. 1987.

Shrub. Miombo woodland. 620 m.

Distribution: SL—*Harder 1509*

Monadenium orobanchoides Bally vel sp. aff.; Carter, F.T.E.A. (Euphorbiaceae 2): 544. 1988.

Terrestrial herb. Montane grassland. 1860–2200 m.

Distribution: SL—*Gereau 3931*

Monadenium pedunculatum S. Carter; Carter, F.T.E.A. (Euphorbiaceae 2): 542. 1988.

Terrestrial herb. Woodland, montane grassland. 1250–1810 m.

Distribution: MP, SL—*Harder 1491*

Neoboutonia macrocalyx Pax; Radcliffe-Smith,

F.T.E.A. (Euphorbiaceae 1): 232. 1987.

Tree. Montane forest. 1960–2300 m.

Distribution: MP, MR, SL—*Mwasumbi 16360*

Phyllanthus amarus Schumach & Thonn.; Radcliffe-Smith, F.T.E.A. (Euphorbiaceae 1): 58. 1987.

Terrestrial herb. Montane forest. 1900 m.

Distribution: NL—*Gereau 4222*

Phyllanthus arvensis Müll. Arg.; Radcliffe-Smith, F.T.E.A. (Euphorbiaceae 1): 57. 1987.

Terrestrial herb. Riverbanks in montane grassland. 2550–2890 m.

Distribution: KP—*Gereau 3115*

Phyllanthus capillaris Schumach. & Thonn.; Hutchinson, F.T.A. 6(1): 709. 1912.

Terrestrial herb. Riverine forest, montane forest. 1210–1790 m.

Distribution: RM, SL—*Gereau 4249*

Phyllanthus micromeris Radcl.-Sm.; Radcliffe-Smith, F.T.E.A. (Euphorbiaceae 1): 59. 1987.

Terrestrial herb. Montane forest. 2300 m.

Distribution: MR—*Gereau 4585*

Phyllanthus muellerianus (Kuntze) Exell; Radcliffe-Smith, F.T.E.A. (Euphorbiaceae 1): 24. 1987.

Shrub. Montane forest. 2075 m.

Distribution: RM—*Semsei 1563*

Phyllanthus niruroides Müll. Arg.; Radcliffe-Smith, F.T.E.A. (Euphorbiaceae 1): 60. 1987.

Terrestrial herb. Miombo woodland. 1200 m.

Distribution: MP—*J. Lovett 4139*

Phyllanthus paxii Hutch.; Radcliffe-Smith, F.T.E.A. (Euphorbiaceae 1): 56. 1987.

Terrestrial herb. Miombo woodland, grassland, riverine forest. 1500–1940 m.

Distribution: SL—*Gereau 4666*

Phyllanthus pseudoniruri Müll. Arg.; Radcliffe-Smith, F.T.E.A. (Euphorbiaceae 1): 48. 1987.

Terrestrial herb. Woodland. 1250 m.

Distribution: MP—*J. Lovett 4074*

Phyllanthus ukagurensis Radcl.-Sm.; Radcliffe-Smith, F.T.E.A. (Euphorbiaceae 1): 44. 1987.

Terrestrial herb. Trailsides in montane forest. 2040 m.

Distribution: SL—*Gereau 4705*

Pseudolachnostylis maprouneifolia Pax var. *dekindtii* (Pax) Radcl.-Sm.; Radcliffe-Smith, F.T.E.A. (Euphorbiaceae 1): 81. 1987.

Tree. Riverine forest in miombo woodland. 1200–1250 m.

Distribution: MP—*J. Lovett 4360*

Pseudolachnostylis maprouneifolia Pax var. *maprouneifolia*; Radcliffe-Smith, F.T.E.A. (Euphorbiaceae 1): 81. 1987.

Tree. Coffee plantation, miombo woodland. 1300–1410 m.

Distribution: MP—*Kayombo 1058*

Pseudolachnostylis maprouneifolia Pax var. *polygyna* (Pax & K. Hoffm.) Radcl.-Sm.; Radcliffe-Smith, F.T.E.A. (Euphorbiaceae 1): 83. 1987.

Tree. Miombo woodland. 1300 m.

Distribution: MP—*J. Lovett 4909*

Ricinus communis L.; Radcliffe-Smith, F.T.E.A. (Euphorbiaceae 1): 322. 1987.

Shrub. Riverine forest, woodland, montane forest. 1310–1980 m.

Distribution: MP, RM—*Kayombo 1109*

Tragia furialis Bojer ex Prain; Radcliffe-Smith, F.T.E.A. (Euphorbiaceae 1): 316. 1987.

Herbaceous climber. Woodland. 1250 m.

Distribution: MP—*J. Lovett 4424*

Tragia plukenetii Radcl.-Sm.; Radcliffe-Smith, F.T.E.A. (Euphorbiaceae 1): 296. 1987.

Terrestrial herb. Riverine forest. 1275 m.

Distribution: MP—*Kayombo 2376*

Tragiella anomala (Prain) Pax & K. Hoffm.; Radcliffe-Smith, F.T.E.A. (Euphorbiaceae 1): 321. 1987.

Herbaceous climber. Montane forest. 2200 m.

Distribution: MR—*Mwasumbi 16202*

Uapaca kirkiana Müll. Arg.; Radcliffe-Smith, F.T.E.A. (Euphorbiaceae 2): 567. 1988.

Tree. Woodland. 1260–1800 m.

Distribution: MP, NL, SL—*Kayombo 2390*

Uapaca nitida Müll. Arg.; Radcliffe-Smith, F.T.E.A. (Euphorbiaceae 2): 568. 1988.

Tree. Woodland. 620–1400 m.

Distribution: MP, SL—*Harder 1524*

FABACEAE

42 genera and 156 species

Abrus melanospermus Hassk. subsp. *tenuiflorus* (Spruce ex Benth.) D. Harder; Harder, Fl. Zamb. 3(3): 216. 2007.

Lianescent shrub. Riverine forest. 580 m.

Distribution: SL—*Gereau 3802*

Adenocarpus mannii (Hook. f.) Hook. f.; Milne-Redhead, F.T.E.A. (Leguminosae–Papilionoideae 2): 1009. 1971.

Shrub. Stream banks, montane grassland and heathland. 1950–2960 m.

Distribution: KP, MP, MR, RM—*Abeid 2877*

Aeschynomene abyssinica (A. Rich.) Vatke; Verdcourt, F.T.E.A. (Leguminosae–Papilionoideae 1): 396. 1971.

Shrublet. Montane grassland. 2300–2610 m.

Distribution: KP, MP—*Pócs 87063*

Aeschynomene mimosifolia Vatke; Verdcourt, F.T.E.A. (Leguminosae–Papilionoidede 1): 392. 1971.

Shrub. Shrubby montane grassland. 2300–2400 m.

Distribution: MP—*P. Lovett 501*

Aeschynomene nyassana Taub.; Verdcourt, F.T.E.A. (Leguminosae–Papilionoideae 1): 403. 1971.

Terrestrial herb. Miombo woodland, rocky scrubland. 1780–2000 m.

Distribution: MP, SL—*Gereau 3908*

Aeschynomene nyikensis Baker; Verdcourt, F.T.E.A. (Leguminosae–Papilionoideae 1): 394. 1971.

Shrub. Riverine forest. 1500 m.

Distribution: RM, SL—*Gereau 4792*

Aeschynomene schimperi Hochst. ex A. Rich.; Verdcourt, F.T.E.A. (Leguminosae–Papilionoideae 1): 376. 1971.

Shrub. Wet grassland. 1300–1400 m.

Distribution: MP—*Kayombo 1043*

Aeschynomene schliebenii Harms var. *mossambicensis* (Baker f.) Verdc.; Verdcourt, F.T.E.A. (Leguminosae–Papilionoideae 1): 395. 1971.

Shrub. Miombo woodland on steep rocky slope. 1850 m.

Distribution: MR—*Carter 2585*

Aeschynomene stolzii Harms; Verdcourt, F.T.E.A. (Leguminosae–Papilionoideae 1): 388. 1971.

Subshrub or perennial herb. Montane grassland. 2500 m.

Distribution: KP, NL, RM—*J. Lovett 776*

Aeschynomene cf. *trigonocarpa* Taub. ex Baker f.; Verdcourt, F.T.E.A. (Leguminosae–Papilionoideae 1): 399. 1971.

Subshrub. Montane grassland. 2570 m.

Distribution: KP—*Gereau 3094*

Alysicarpus glumaceus (Vahl) DC. subsp. *glumaceus*; Verdcourt, F.T.E.A. (Leguminosae–Papilionoideae 1): 498. 1971.

Terrestrial herb. Abandoned cultivation. 1250 m.

Distribution: MP—*J. Lovett 4077*

Alysicarpus rugosus (Willd.) DC. subsp. *perennirufus* J. Léonard; Verdcourt, F.T.E.A. (Leguminosae–Papilionoideae 1): 496. 1971.

Terrestrial herb. Grassland. 1250 m.

Distribution: MP—*J. Lovett 4443*

Amphicarpaea africana (Hook. f.) Harms; Verdcourt, F.T.E.A. (Leguminosae–Papilionoideae 2): 511. 1971.

Herbaceous climber. Montane forest. 2080 m.

Distribution: MR—*Mwasumbi 16192*

Argyrolobium macrophyllum Harms; Polhill, F.T.E.A. (Leguminosae–Papilionoideae 2): 1002. 1971.

Terrestrial herb. Roadsides in montane grassland. 1940–2140 m.

Distribution: SL—*Suleiman 145*

Argyrolobium rupestre (E. Mey.) Walp. subsp. *aberdaricum* (Harms) Polhill; Polhill, F.T.E.A. (Leguminosae–Papilionoideae 2): 1006. 1971.

Terrestrial herb. Montane grassland. 2320–2740 m.

Distribution: KP, NL, SL—*Gereau 4095*

Argyrolobium stolzii Harms; Polhill, F.T.E.A. (Leguminosae–Papilionoideae 2): 1006. 1971.

Terrestrial herb. Montane heathland. 2620–2770 m.

Distribution: RM—*Mwasumbi 16263*

Argyrolobium tomentosum (Andrews) Druce; Polhill, F.T.E.A. (Leguminosae–Papilionoideae 2): 1007. 1971.

Terrestrial herb or shrub. Montane grassland and bushland. 1560–2300 m.

Distribution: MP, RM, SL—*Mwasumbi 16277*

Argyrolobium vaginiferum Harms; Polhill, F.T.E.A. (Leguminosae–Papilionoideae 2): 1008. 1971.

Terrestrial herb or subshrub. Montane grassland and marsh. 2500–2840 m.

Distribution: KP, NL, RM—*Goetze 1134* (type: K), *Abeid 2783*

Astragalus atropilosulus (Hochst.) Bunge subsp. *burkeanus* (Benth. ex Harv.) J. B. Gillett; Gillett, F.T.E.A. (Leguminosae–Papilionoideae 2): 1056. 1971.

Shrub or perennial herb. Montane grassland, edges of cultivated fields. 2200–2840 m.

Distribution: KP, MP, MR—*Suleiman 40*

Centrosema pubescens Benth.; Verdcourt, F.T.E.A. (Leguminosae–Papilionoideae 2): 520. 1971.

Herbaceous climber. Grassland. 1300–1400 m.

Distribution: MP—*Kayombo 974*

Craibia brevicaudata (Vatke) Dunn subsp. *baptistarum* (Büttner) J. B. Gillett; Gillett, F.T.E.A. (Leguminosae–Papilionoideae 1): 151. 1971.

Tree. Riverine forest, montane forest. 1500–1950 m.

Distribution: MP, RM, SL—*J. Lovett 4814*

Crotalaria aculeata De Wild. subsp. *aculeata*; Polhill, Fl. Zamb. 3(7): 186. 2003.

Shrublet. Grassland. 1250–1300 m.

Distribution: MP—*J. Lovett 4236*

Crotalaria adolfi Harms; Polhill, F.T.E.A. (Leguminosae–Papilionoideae 2): 861. 1971.

Shrub. Riverine forest, montane forest. 900–2000 m.

Distribution: NL, SL—*Stolz 2238* (type: BM, MO, P, Z), *Richards 9856*

Crotalaria agatiflora Schweinf. subsp. *vaginifera* Polhill; Polhill, F.T.E.A. (Leguminosae–Papilionoideae 2): 853. 1971.

Shrub. Riverine forest, montane forest edges. 1210–2590 m.

Distribution: MP, SL—*Procter 1513* (type: EA, K), *Gereau 4247*

Crotalaria alexandri Baker f.; Polhill, F.T.E.A. (Leguminosae–Papilionoideae 2): 992. 1971.

Terrestrial herb. Woodland. 1250 m.

Distribution: MP—*J. Lovett 4504*

Crotalaria anthyllopsis Welw. ex Baker; Polhill, F.T.E.A. (Leguminosae–Papilionoideae 2): 927. 1971.

Terrestrial herb. Abandoned cultivation. 1300–1400 m.

Distribution: MP—*Kayombo 1018*

Crotalaria argyrolobioides Baker; Polhill, F.T.E.A. (Leguminosae–Papilionoideae 2): 975. 1971.

Terrestrial herb. Miombo woodland. 1600 m.

Distribution: MP—*P. Lovett 234*

Crotalaria becquetii Baker f. ex R. Wilczek subsp. *becquetii*; Polhill, F.T.E.A. (Leguminosae–Papilionoideae 2): 856. 1971.

Shrub. Montane forest and shrubland, riverine forest. 1500–2240 m.

Distribution: MP, MR, NL, RM—*Gereau 4198*

Crotalaria brevidens Benth. var. *intermedia* (Kotschy) Polhill; Polhill, F.T.E.A. (Leguminosae–Papilionoideae 2): 909. 1971.

Shrub. Montane grassland. 2280 m.

Distribution: MR—*Mwasumbi 16169*

Crotalaria cephalotes Steud. ex A. Rich.; Polhill, F.T.E.A. (Leguminosae–Papilionoideae 2): 983. 1971.

Terrestrial herb. Woodland, grassland. 1100–1400 m.

Distribution: MP—*Kayombo 1013*

Crotalaria chirindae Baker f.; Polhill, F.T.E.A. (Leguminosae–Papilionoideae 2): 923. 1971.

Shrub. Edges of cultivated fields. 2200 m.

Distribution: MR—*Gereau 4502*

Crotalaria cleomifolia Welw. ex Baker; Polhill, F.T.E.A. (Leguminosae–Papilionoideae 2): 922. 1971.

Shrub or perennial herb. Montane forest edges and clearings. 1900–2140 m.

Distribution: MR, NL, RM, SL—*Suleiman 121*

Crotalaria hyssopifolia Klotzsch; Polhill, F.T.E.A. (Leguminosae–Papilionoideae 2): 991. 1971.

Terrestrial herb. Miombo woodland. 1600–1800 m.

Distribution: MP—*P. Lovett 236*

Crotalaria incana L. subsp. *purpurascens* (Lam.) Milne-Redh.; Polhill, F.T.E.A. (Leguminosae–

Papilionoidede 2): 870. 1971.

Terrestrial herb. Cultivated fields, waste ground. 1250–2100 m.

Distribution: MP, MR—*Kayombo 2464*

Crotalaria lachnocarpoides Engl.; Polhill, F.T.E.A. (Leguminosae–Papilionoideae 2): 879. 1971.

Shrub or perennial herb. Montane forest edges, grassland, edges of cultivation. 1250–2250 m.

Distribution: MP, MR, RM, SL—*Suleiman 147*

Crotalaria natalitia Meisn. var. *natalitia*; Polhill, F.T.E.A. (Leguminosae–Papilionoideae 2): 872. 1971.

Shrub. Woodland, riverine forest. 1250–1710 m.

Distribution: MP, RM, SL—*Gereau 4798*

Crotalaria natalitia Meisn. var. *rutshuruensis* De Wild.; Polhill, F.T.E.A. (Leguminosae–Papilionoideae 2): 872. 1971.

Shrub. Montane forest edges and bushland. 2040–2300 m.

Distribution: MP, MR, SL—*Gereau 4749*

Crotalaria cf. *nudiflora* Polhill; Polhill, Fl. Zamb. 3(7): 151. 2003.

Terrestrial herb. Miombo woodland. 1780 m.

Distribution: SL—*Gereau 5105*

Crotalaria ochroleuca G. Don; Polhill, F.T.E.A. (Leguminosae–Papilionoideae 2): 908. 1971.

Terrestrial herb. Woodland, abandoned cultivation. 1250 m.

Distribution: MP—*J. Lovett 4648*

Crotalaria oocarpa Baker subsp. *oocarpa*; Polhill, F.T.E.A. (Leguminosae–Papilionoideae 2): 968. 1971.

Terrestrial herb. Roadsides in miombo woodland, disturbed riverine forest. 1150–1500 m.

Distribution: MP, SL—*Kayombo 2454*

Crotalaria recta Steud. ex A. Rich.; Polhill, F.T.E.A. (Leguminosae–Papilionoideae 2): 957. 1971.

Terrestrial herb. Riverine forest, montane forest margins, wet grassland, woodland, cultivated fields. 1250–2000 m.

Distribution: MP, RM—*Kayombo 2377*

Crotalaria ringoetii Baker f.; Polhill, F.T.E.A. (Leguminosae–Papilionoideae 2): 987. 1971.

Terrestrial herb. Miombo woodland. 1800 m.

Distribution: MP—*P. Lovett 51*

Crotalaria rogersii Baker f.; Polhill, F.T.E.A. (Leguminosae–Papilionoideae 2): 906. 1971.

Terrestrial herb. Lakeshore sand. 480 m.

Distribution: SL—*Gereau 3659*

Crotalaria seemeniana Harms; Polhill, F.T.E.A. (Leguminosae–Papilionoideae 2): 975. 1971.

Terrestrial herb. Montane grassland, roadsides, tree plantations, cultivated ground. 1500–2290 m.

Distribution: MR, RM—*Greenway 3581*

Crotalaria senegalensis (Pers.) Bacle ex DC.; Polhill, F.T.E.A. (Leguminosae–Papilionoideae 2): 967. 1971.

Terrestrial herb. Cultivated fields. 1250 m.

Distribution: MP—*J. Lovett 4623*

Crotalaria shirensis (Baker f.) Milne-Redh.; Polhill, F.T.E.A. (Leguminosae–Papilionoideae 2): 946. 1971.

Terrestrial herb. Miombo woodland. 1800 m.

Distribution: MP—*P. Lovett 243*

Crotalaria sparsifolia Baker; Polhill, F.T.E.A. (Leguminosae–Papilionoideae 2): 988. 1971.

Terrestrial herb. Miombo woodland. 1475 m.

Distribution: MP—*Kayombo 2393*

Crotalaria spartea R. Br. ex Baker; Polhill, F.T.E.A. (Leguminosae–Papilionoideae 2): 937. 1971.

Terrestrial herb. Grassland, abandoned cultivation. 1250–1500 m.

Distribution: MP, SL—*Gereau 4405*

Crotalaria sphaerocarpa Perr. ex DC.; Polhill, F.T.E.A. (Leguminosae–Papilionideae 2): 942. 1971.

Terrestrial herb. Open woodland. 1650–1680 m.

Distribution: SL—*Gereau 5111*

Crotalaria subcapitata De Wild. subsp. *subcapitata*; Polhill, 1982: 197.

Terrestrial herb. Montane grassland. 1940 m.

Distribution: SL—*Gereau 4667*

Crotalaria teretifolia Milne-Redh.; Polhill, F.T.E.A. (Leguminosae–Papilionoideae 2): 946. 1971.

Terrestrial herb. Grassland. 1250 m.

Distribution: MP—*J. Lovett 4390*

Crotalaria valida Baker; Polhill, F.T.E.A. (Leguminosae–Papilionoideae 2): 879. 1971.

Terrestrial herb. Montane grassland. 1810–1860 m.

Distribution: KP, SL—*Gereau 3923*

Dalbergia arbutifolia Baker subsp. *arbutifolia*; Polhill, F.T.E.A. (Leguminosae–Papilionoideae 1): 108. 1971.

Liana. Riverine forest, lakeshore thickets. 474–1400 m.

Distribution: MP, SL—*Gereau 3674*

Dalbergia fischeri Taub.; Polhill, F.T.E.A. (Leguminosae–Papilionoideae 1): 110. 1971.

Liana or scandent shrub. Riverine forest, lakeshore woodland. 474–1210 m.

Distribution: MP, SL—*Mwasumbi 16589*

Dalbergia lactea Vatke; Polhill, F.T.E.A. (Leguminosae–Papilionoideae 1): 111. 1971.

Liana or shrub. Montane forest, riverine forest. 1290–2100 m.

Distribution: MP, RM, SL—*Gereau 3613*

Desmodium adscendens (Sw.) DC.; Schubert, F.T.E.A. (Leguminosae–Papilionoideae 1): 461. 1971.

Terrestrial herb. Grassland. 1360 m.

Distribution: RM—*Hooper 845*

Desmodium appressipilum B. G. Schub.; Schubert, F.T.E.A. (Leguminosae–Papilionoideae 1): 458. 1971.

Terrestrial herb. Woodland, montane grassland. 1250–2100 m.

Distribution: KP, MP—*Richards* 14184 (type: A, K), *J. Lovett 4505*

Desmodium barbatum (L.) Benth. var. *argyreum* (Welw. ex Baker) B. G. Schub.; Schubert, F.T.E.A. (Leguminosae–Papilionoideae 1): 479. 1971.

Woody-based herb. Woodland. 1400–1500 m.

Distribution: MP—*J. Lovett 3400*

Desmodium repandum (Vahl) DC.; Schubert, F.T.E.A. (Leguminosae–Papilionoideae 1): 465. 1971.

Terrestrial herb, often scrambling. Montane grassland, montane forest. 1780–2360 m.

Distribution: MR, NL, RM, SL—*Suleiman 102*

Desmodium salicifolium (Poir.) DC.; Schubert, F.T.E.A. (Leguminosae–Papilionoideae 1): 469. 1971.

Shrub. Riverine forest, edges of rice fields. 600–1210 m.

Distribution: MR, SL—*Gereau 4251*

Desmodium tortuosum (Sw.) DC.; Schubert, F.T.E.A. (Leguminosae–Papilionoideae 1): 474. 1971.

Terrestrial herb. Lakeshore forest. 474 m.

Distribution: SL—*Gereau 3672*

Dolichos kilimandscharicus Taub. subsp. *kilimandscharicus*; Verdcourt, F.T.E.A. (Leguminosae–Papilionoideae 2): 685. 1971.

Terrestrial herb. Montane grassland, montane forest edges, woodland. 1300–2250 m.

Distribution: MP, RM, SL—*Gereau 3583*

Dolichos sericeus E. Mey. subsp. *sericeus*; Verdcourt, F.T.E.A. (Leguminosae–Papilionoideae 2): 680. 1971.

Herbaceous climber. Montane forest edges, shrubland. 2180–2400 m.

Distribution: MR—*Mwasumbi 16175*

Dolichos simplicifolius Hook. f.; Verdcourt, F.T.E.A. (Leguminosae–Papilionoideae 2): 692. 1971.

Terrestrial herb. Seasonally inundated grassland, woodland. 1270–1310 m.

Distribution: MP—*Gereau 3478*

Dolichos trinervatus Baker; Verdcourt, F.T.E.A. (Leguminosae–Papilionoideae 2): 688. 1971.

Terrestrial herb. Miombo woodland. 1200–1500 m.

Distribution: MP—*Gereau 7001*

Dolichos xiphophyllus Baker; Verdcourt, F.T.E.A. (Leguminosae–Papilionoideae 2): 691. 1971.

Terrestrial herb. Grassland. 1250 m.

Distribution: MP—*Lovett 3645*

Dumasia villosa DC.; Verdcourt, F.T.E.A. (Leguminosae–Papilionoideae 2): 513. 1971.

Herbaceous climber. Montane forest. 1790–2100 m.

Distribution: MR, RM—*Mwasumbi 16310*

Eriosema buchananii Baker f. var. *buchananii*; Verdcourt, F.T.E.A. (Leguminosae–Papilionoideae 2): 776. 1971.

Terrestrial herb. Montane grassland. 2120 m.

Distribution: SL—*Gereau 4329*

Eriosema buchananii Baker f. var. *subprostratum* Verdc.; Verdcourt, F.T.E.A. (Leguminosae–Papilionoideae 2): 777. 1971.

Terrestrial herb. Montane grassland. 2250–2840 m.

Distribution: KP, MR, RM—*Abeid 2780*

Eriosema ellipticum Welw. ex Baker; Verdcourt, F.T.E.A. (Leguminosae–Papilionoideae 2): 768. 1971.

Shrub or subshrub. Woodland, wooded grassland. 900–1400 m.

Distribution: MP, RM, SL—*Kibure 1483*

Eriosema flexuosum Staner; Verdcourt, F.T.E.A. (Leguminosae–Papilionoideae 2): 804. 1971.

Terrestrial herb. Montane grassland. 2500 m.

Distribution: KP—*Abeid 3110*

Eriosema lebrunii Staner & De Craene; Verdcourt, F.T.E.A. (Leguminosae–Papilionoideae 2): 802. 1971.

Terrestrial herb. Montane grassland. 2120–2488 m.

Distribution: SL—*Suleiman 181*

Eriosema macrostipulum Baker f. var. *macrostipulum*; Verdcourt, F.T.E.A. (Leguminosae–Papilionoideae 2): 789. 1971.

Terrestrial herb. Woodland. 1200 m.

Distribution: MP—*J. Lovett 3867*

Eriosema montanum Baker f. var. *brevipedunculatum*; Verdcourt, F.T.E.A. (Leguminosae–Papilionoideae 2): 78. 1971.

Shrub. Edges of cultivated fields. 2200 m.

Distribution: MR—*Gereau 4509*

Eriosema montanum Baker f. var. *montanum*; Verdcourt, F.T.E.A. (Leguminosae–Papilionoideae 2): 780. 1971.

Shrub. Montane forest, forest edges. 2040–2100 m.

Distribution: MR, SL—*Gereau 4751*

Eriosema nutans Schinz; Verdcourt, F.T.E.A. (Leguminosae–Papilionoideae 2): 775. 1971.

Terrestrial herb. Montane grassland. 2230–2360 m.

Distribution: MR, SL—*Gereau 3773*

Eriosema parviflorum E. Mey. subsp. *parviflorum*; Verdcourt, F.T.E.A. (Leguminosae–Papilionoideae 2): 779. 1971.

Terrestrial herb. Grassland.

Distribution: RM—*Eggeling 6518*

Eriosema shirense Baker f.; Verdcourt, F.T.E.A. (Leguminosae–Papilionoideae 2): 798. 1971.

Terrestrial herb. Woodland. 1200–1790 m.

Distribution: MP, SL—*Gereau 4381*

Eriosema ukingense Harms; Verdcourt, F.T.E.A. (Leguminosae–Papilionoideae 2): 777. 1971.

Terrestrial herb. Rocky mountain slopes. 2800 m.

Distribution: KP, MP—*Goldblatt 8207*

Eriosema verdickii De Wild.; Verdcourt, F.T.E.A. (Leguminosae–Papilionoideae 2): 803. 1971.

Terrestrial herb. Woodland. 1250–1790 m.

Distribution: MP, SL—*Gereau 3893*

Erythrina abyssinica Lam.; Verdcourt, F.T.E.A. (Leguminosae–Papilionoideae 2): 555. 1971.

Tree. Montane forest remnants, forest edges, riverine forest, woodland, wooded grassland. 1300–1840 m.

Distribution: MP, SL—*Gereau 3547*

Flemingia grahamiana Wight & Arn.; Verdcourt, F.T.E.A. (Leguminosae–Papilionoideae 2): 806. 1971.

Terrestrial herb. Wooded grassland. 980 m.

Distribution: SL—*Kibure 1485*

Humularia drepanocephala (Baker) P. A. Duvign. var. *drepanocephala*; Verdcourt, F.T.E.A. (Leguminosae–Papilionoideae 1): 431. 1971.

Terrestrial herb. Woodland. 1400–2250 m.

Distribution: MP—*J. Lovett 4853*

Indigofera arrecta Hochst. ex. A. Rich.; Gillett, F.T.E.A. (Leguminosae–Papilionoideae 1): 307. 1971.

Woody-based herb. Woodland, montane grassland. 1350–2520 m.

Distribution: KP, MP—*J. Lovett 4691*

Indigofera asterocalycina Gilli; Gillett, F.T.E.A. (Leguminosae–Papilionoideae 1): 276. 1971.

Shrub. Cultivated cassava fields. 474–480 m.

Distribution: SL—*Mwasumbi 16575*

Indigofera astragalina DC.; Gillett, F.T.E.A. (Leguminosae–Papilionoideae 1): 310. 1971.

Terrestrial herb. Lakeshore shrubland, abandoned cultivation. 474–1250 m.

Distribution: MP, SL—*Mwasumbi 16576*

Indigofera atricephala J. B. Gillett; Gillett, F.T.E.A. (Leguminosae–Papilionoideae 1): 284. 1971.

Terrestrial herb. Disturbed montane forest. 2280–2400 m.

Distribution: RM—*Mwasumbi 16243*

Indigofera atriceps Hook. f. subsp. *atriceps*; Gillett, F.T.E.A. (Leguminosae–Papilionoideae 1): 282. 1971.

Shrub. Miombo woodland, montane forest edges, riverine forest. 1500–2400 m.

Distribution: MR, RM, SL—*Gereau 4501*

Indigofera atriceps Hook. f. subsp. *glandulosissima* (R. E. Fr.) J. B. Gillett; Gillett, F.T.E.A. (Leguminosae–Papilionoideae 1): 282. 1971.

Woody-based herb. Miombo woodland. 1300–1800 m.

Distribution: MP—*P. Lovett 60*

Indigofera atriceps Hook. f. vel sp. aff., Taxon "B"; Gillett, F.T.E.A. (Leguminosae–Papilionoideae 1): 284. 1971.

Terrestrial herb. Montane grassland. 2100–2600 m.

Distribution: KP—*J. Lovett 1854*

Indigofera demissa Taub.; Gillett, F.T.E.A. (Leguminosae–Papilionoideae 1): 225. 1971.

Terrestrial herb. Abandoned cultivation. 480–900 m.

Distribution: NL—*Pócs 87649*

Indigofera dendroides Jacq.; Gillett, F.T.E.A. (Leguminosae–Papilionoideae 1): 242. 1971.

Woody-based terrestrial herb. Wooded grassland, montane thickets. 1400–2400 m.

Distribution: MP, MR—*Kayombo 2356*

Indigofera emarginella Steud. ex A. Rich.; Gillett, F.T.E.A. (Leguminosae–Papilionoideae 1): 300. 1971.

Shrublet. Miombo woodland. 1200–1790 m.

Distribution: MP, SL—*Kayombo 2424*

Indigofera fulvopilosa Brenan; Gillett, F.T.E.A. (Leguminosae–Papilionideae 1): 257. 1971.

Terrestrial herb. Premontane shrubland. 1610 m.

Distribution: RM—*Mwasumbi 16344*

Indigofera fuscosetosa Baker; Gillett, F.T.E.A. (Leguminosae–Papilionoideae 1): 279. 1971.

Terrestrial herb. Montane grassland, premontane shrubland. 1610–1940 m.

Distribution: RM, SL—*Mwasumbi 16339*

Indigofera homblei Baker f. & Martin subsp. *homblei*; Gillett, F.T.E.A. (Leguminosae–Papilionoideae 1): 289. 1971.

Shrub. Miombo woodland, riverine forest edges. 1475–1800 m.

Distribution: MP, SL—*Kayombo 2391*

Indigofera masaiensis J. B. Gillett; Gillett, F.T.E.A. (Leguminosae–Papilionoideae 1): 264. 1971.

Scrambling terrestrial herb. Bushy montane grassland. 2100 m.

Distribution: MP—*P. Lovett 406*

Indigofera mimosoides Baker; Gillett, F.T.E.A. (Leguminosae–Papilionoideae 1): 262. 1971.

Terrestrial herb. Montane shrubland. 1760–2540 m.

Distribution: KP, MP, NL—*Abeid 2871*

Indigofera paniculata Vahl ex Pers. subsp. *paniculata*; Gillett, F.T.E.A. (Leguminosae–Papilionoideae 1): 232. 1971.

Shrub. Riverine forest edges in miombo woodland. 1500 m.

Distribution: SL—*Gereau 4411*

Indigofera rhynchocarpa Welw. ex Baker var. *rhynchocarpa*; Gillett, F.T.E.A. (Leguminosae–Papilionoideae 1): 297. 1971.

Shrub. Miombo woodland, field edges, and abandoned cultivation. 1250–1800 m.

Distribution: MP—*P. Lovett 250*

Indigofera roseocaerulea Baker f.; Gillett, F.T.E.A. (Leguminosae–Papilionoideae 1): 299. 1971.

Shrub. Montane forest edges, bushy montane grassland. 1780–2230 m.

Distribution: MP, MR—*Mwasumbi 16174*

Indigofera schliebenii Harms; Gillett, F.T.E.A. (Leguminosae–Papilionoideae 1): 284. 1971.

Terrestrial herb. Montane grassland. 1940 m.

Distribution: SL—*Gereau 4687*

Indigofera subulifera Welw. ex Baker; Gillett, F.T.E.A. (Leguminosae–Papilionoideae 1): 274. 1971.

Terrestrial herb. Miombo woodland. 1790 m.

Distribution: SL—*Gereau 3904*

Indigofera trachyphylla Benth. ex Oliv.; Gillett, F.T.E.A. (Leguminosae–Papilionoideae 1): 278. 1971.

Terrestrial herb. Miombo woodland. 1530 m.

Distribution: SL—*Gereau 4373*

Indigofera ugandensis Baker f.; Gillett, F.T.E.A. (Leguminosae–Papilionoideae 1): 273. 1971.

Terrestrial herb. Montane grassland. 1920 m.

Distribution: SL—*Gereau 4671*

Indigofera zenkeri Harms ex Baker f.; Gillett, F.T.E.A. (Leguminosae–Papilionoideae 1): 265. 1971.

Terrestrial herb. Lakeshore shrubland. 474–480 m.

Distribution: SL—*Mwasumbi 16580*

Kotschya africana Endl.; Verdcourt, F.T.E.A. (Leguminosae–Papilionoideae 1): 412. 1971.

Shrub. Riverine forest. 1500 m.

Distribution: SL—*Gereau 4780*

Kotschya coalescens Dewit & P. A. Duvign.; Verdcourt, F.T.E.A. (Leguminosae–Papilionoideae 1): 425. 1971.

Shrub. Shrubby grassland. 2600 m.

Distribution: MP, MR—*Iwarsson 1187*

Kotschya recurvifolia (Taub.) F. White subsp. *recurvifolia*; Verdcourt, F.T.E.A. (Leguminosae–Papilionoideae 1): 417. 1971.

Shrub or tree. Montane heathland, forest edges, wooded grassland. 2100–2770 m.

Distribution: MP, RM, SL—*Mwasumbi 16260*

Kotschya strigosa (Benth.) Dewit & P. A. Duvign. var. *strigosa*; Verdcourt, F.T.E.A. (Leguminosae–Papilionoideae 1): 423. 1971.

Shrub or tree. Woodland, montane grassland, bamboo forest. 1400–2200 m.

Distribution: MP, RM, SL—*Gereau 4689*

Kotschya thymodora (Baker f.) Wild subsp. *septentrionalis* Verdc.; Verdcourt, F.T.E.A. (Leguminosae–Papilionoideae 1): 416. 1971.

Shrub. Montane grassland and heathland, bushed grassland. 1940–2960 m.

Distribution: KP, RM, SL—*Abeid 2867*

Lonchocarpus capassa Rolfe; Polhill, F.T.E.A. (Leguminosae–Papilionoideae 1): 66. 1971.

Tree. Miombo woodland, lakeshore forest. 480–1300 m.

Distribution: MP, SL—*Gereau 3651*

Lotononis angolensis Welw. ex Baker; Milne-Redhead, F.T.E.A. (Leguminosae–Papilionoideae 2): 814. 1971.

Terrestrial herb. Montane grassland, roadsides in miombo woodland. 1500–1940 m.

Distribution: RM, SL—*Mwasumbi 16342*

Lotononis stolzii Harms; Milne-Redhead, F.T.E.A. (Leguminosae–Papilionoideae 2): 816. 1971.

Terrestrial herb. Montane grassland. 2300–2600 m.

Distribution: KP, NL, SL—*Stolz 2602* (type: BM, BOL, C, EA, K, MO, PRE), *J. Lovett 1855*

Lotus discolor E. Mey.; Gillett, F.T.E.A. (Leguminosae–Papilionoideae 2): 1047. 1971.

Terrestrial herb. Woodland, montane grassland. 1400–2740 m.

Distribution: KP, MP—*J. Lovett 3398*

Lotus subdigitatus Boutique; Gillett, F.T.E.A. (Leguminosae–Papilionoideae 2): 1047. 1971.

Terrestrial herb. Montane grassland. 2500 m.

Distribution: RM—*Mwasumbi 16252*

Macrotyloma axillare (E. Mey.) Verdc.; Verdcourt, F.T.E.A. (Leguminosae–Papilionoideae 2): 586. 1971.

Herbaceous climber. Miombo woodland. 1780 m.

Distribution: SL—*Gereau 5094*

Macrotyloma ellipticum (R. E. Fr.) Verdc.; Verdcourt, F.T.E.A. (Leguminosae–Papilionoideae 2): 589. 1971.

Terrestrial herb. Woodland. 1600 m.

Distribution: RM—*Stolz 1335*

Macrotyloma fimbriatum (Harms) Verdc.; Verdcourt, F.T.E.A. (Leguminosae–Papilionoideae 2): 591. 1971.

Terrestrial herb. Miombo woodland. 1530 m.

Distribution: SL—*Gereau 4369*

Microcharis ephemera (J. B. Gillet) Schrire; Schrire, 1992: 166.

Terrestrial herb. Woodland. 1300–1600 m.

Distribution: MP—*P. Lovett 99*

Millettia oblata Dunn subsp. *intermedia* J. B.

Gillett; Gillett, F.T.E.A. (Leguminosae–Papilionoideae 1): 143. 1971.

Tree. Montane forest. 1780 m.

Distribution: RM—*Bracebridge 29*

Millettia oblata Dunn subsp. *stolzii* J. B. Gillett; Gillett, F.T.E.A. (Leguminosae–Papilionoideae 1): 143. 1971.

Tree. Montane forest. 1400–1500 m.

Distribution: RM—*Stolz 183* (type: K, MO), *Davies 301*

Mucuna pruriens (L.) DC. var. *pruriens*; Verdcourt, F.T.E.A. (Leguminosae–Papilionoideae 2): 567. 1971.

Herbaceous climber. Riverine forest. 1250 m.

Distribution: MP—*J. Lovett 4481*

Neonotonia wightii (Graham ex Wight & Arn.) J. A. Lackey subsp. *wightii*; Mackinder, Fl. Zamb. 3(5): 56. 2001.

Herbaceous climber. Roadsides in cultivation. 2200 m.

Distribution: MR—*Gereau 4505*

Neorautanenia mitis (A. Rich.) Verdc.; Verdcourt, F.T.E.A. (Leguminosae–Papilionoideae 2): 700. 1971.

Terrestrial herb, sometimes climbing. Woodland, riverine forest edges. 1250–1600 m.

Distribution: MP, SL—*J. Lovett 4288*

Ormocarpum kirkii S. Moore; Gillett, F.T.E.A. (Leguminosae–Papilionoideae 1): 362. 1971.

Shrub or small tree. Wooded grassland, roadsides. 490–880 m.

Distribution: SL—*Kibure 1487*

Otholobium foliosum (Oliv.) C. H. Stirt.; Verdcourt, Fl. Zamb. 3(6): 46. 2000.

Shrub. Bamboo forest edges, montane grassland. 2270–2360 m.

Distribution: KP, NL—*Bracebridge 111*

Pericopsis angolensis (Baker) Meeuwen; Polhill, F.T.E.A. (Leguminosae–Papilionoideae 1): 41. 1971.

Tree. Woodland. 1200–1400 m.

Distribution: MP—*Kayombo 2414*

Pseudarthria hookeri Wight & Arn. var. *hookeri*; Verdcourt, F.T.E.A. (Leguminosae–Papilionoideae 1): 484. 1971.

Shrub or perennial herb. Woodland, montane grassland, riverine forest. 1200–2440 m.

Distribution: MP, MR, SL—*Suleiman 188*

Pterocarpus angolensis DC.; Polhill, F.T.E.A. (Leguminosae–Papilionoideae 1): 89. 1971.

Tree. Miombo woodland. 1200–1300 m.

Distribution: MP—*J. Lovett 4908*

Pterocarpus tinctorius Welw.; Polhill, F.T.E.A. (Leguminosae–Papilionoideae 1): 86. 1971.

Tree. Miombo woodland. 1200 m.

Distribution: MP—*Kayombo 2438*

Rhynchosia caribaea (Jacq.) DC. var. *caribaea*; Verdcourt, Fl. Zamb. 3(5): 195. 2001.

Herbaceous climber. Riverine forest. 1100 m.

Distribution: MP—*Mwangoka 664*

Note: New to East Africa, from Malawi, Mozambique, Zimbabwe, and South Africa.

Rhynchosia clivorum S. Moore subsp. *pycnantha* (Harms) Verdc.; Verdcourt, Fl. Zamb. 3(5): 179. 2001.

Shrub. Montane grassland and bushland. 2100–2550 m.

Distribution: MP, RM—*Kerfoot 4117*

Rhynchosia goetzei Harms var. *goetzei*; Verdcourt, F.T.E.A. (Leguminosae–Papilionoideae 2): 744. 1971.

Shrub. Woodland. 1400–1500 m.

Distribution: MP—*J. Lovett 3384*

Rhynchosia luteola (Hiern) K. Schum. var. *luteola*; Verdcourt, F.T.E.A. (Leguminosae–Papilionoideae 2): 732. 1971.

Herbaceous climber. Riverine forest. 1300 m.

Distribution: MP—*J. Lovett 4779*

Rhynchosia nyikensis Baker; Verdcourt, F.T.E.A. (Leguminosae–Papilionoideae 2): 729. 1971.

Herbaceous climber. Montane forest. 2040 m.

Distribution: SL—*Gereau 4738*

Sesbania macrantha Welw. ex E. Phillips & Hutch. var. *macrantha*; Gillett, F.T.E.A. (Leguminosae–Papilionoideae 1): 341. 1971.

Shrub. Wooded grassland, woodland, cultivated fields. 1250–1980 m.

Distribution: MP—*Kayombo 2463*

Sesbania sesban (L.) Merr. var. *nubica* Chiov.; Gillett, F.T.E.A. (Leguminosae–Papilionoideae 1): 339. 1971.

Shrub. Sandy lakeshore. 474–490 m.

Distribution: SL—*Mwasumbi 16562*

Sphenostylis stenocarpa (Hochst. ex A. Rich.) Harms; Verdcourt, F.T.E.A. (Leguminosae–Papilionoideae 2): 672. 1971.

Herbaceous climber. Miombo woodland. 1530 m.

Distribution: SL—*Gereau 4355*

Tephrosia aequilata Baker; Gillett, F.T.E.A. (Leguminosae–Papilionoideae 1): 207. 1971.

Shrub. Montane forest. 1760 m.

Distribution: RM—*WCS SHCP 260*

Tephrosia decora Welw. ex Baker; Gillett, F.T.E.A. (Leguminosae–Papilionoideae 1): 180. 1971.

Terrestrial herb. Woodland. 1200 m.

Distribution: MP—*Kayombo 2415*

Tephrosia interrupta Hochst. & Steud. ex Engl. subsp. *elongatiflora* Harms ex J. B. Gillett; Gillett, F.T.E.A. (Leguminosae–Papilionoideae 1): 206. 1971.

Shrub. Montane forest edges. 2110 m.

Distribution: MR—*Gereau 4550*

Tephrosia interrupta Hochst. & Steud. ex Engl. subsp. *mildbraedii* (Harms) J. B. Gillett; Gillett, F.T.E.A. (Leguminosae–Papilionoideae 1): 206. 1971.

Shrub. Edges of cultivated fields. 2200 m.

Distribution: MR—*Gereau 4538*

Tephrosia nyikensis Baker; Gillett, F.T.E.A. (Leguminosae–Papilionoideae 1): 203. 1971.

Shrub. Miombo woodland. 1700 m.

Distribution: MP—*Kayombo 1113*

Tephrosia paniculata Welw. ex Baker; Gillett, F.T.E.A. (Leguminosae–Papilionoideae 1): 176. 1971.

Shrub. Stream banks, edges of cultivated fields. 1500–2200 m.

Distribution: MP, MR, RM—*Mwasumbi 16276*

Tephrosia paucijuga Harms; Gillett, F.T.E.A. (Leguminosae–Papilionoideae 1): 195. 1971.

Terrestrial herb. Montane grassland. 2440 m.

Distribution: SL—*Suleiman 187*

Tephrosia purpurea (L.) Pers. subsp. *leptostachya* (DC.) Brummitt; Gillett, F.T.E.A. (Leguminosae–Papilionoideae 1): 186. 1971.

Terrestrial herb. Grassland. 1250 m.

Distribution: MP—*J. Lovett 4231*

Tephrosia richardsiae J. B. Gillett; Gillett, F.T.E.A. (Leguminosae–Papilionoideae 1): 178. 1971.

Terrestrial herb. Woodland. 1250 m.

Distribution: MP—*J. Lovett 4340*

Tephrosia stormsii De Wild. var. *pilosa* Brummitt; Gillett, F.T.E.A. (Leguminosae–Papilionoideae 1): 198. 1971.

Terrestrial herb. Roadsides. 474–480 m.

Distribution: SL—*Mwasumbi 16572*

Tephrosia vogelii Hook. f.; Gillett, F.T.E.A. (Leguminosae–Papilionoideae 1): 210. 1971.

Terrestrial herb or soft-wooded shrub. Montane forest edges. 2100 m.

Distribution: RM—*WCS SHCP 178*

Teramnus uncinatus (L.) Sw. subsp. *axilliflorus* (Kotschy) Verdc.; Verdcourt, F.T.E.A. (Leguminosae–Papilionoideae 2): 540. 1971.

Herbaceous climber. Woodland. 1300–1400 m.

Distribution: MP—*Kayombo 997*

Trifolium repens L.; Gillett, F.T.E.A. (Leguminosae–Papilionoideae 2): 1016. 1971.

Terrestrial herb. Montane grassland. 2400 m.

Distribution: KP—*Suleiman 42*

Trifolium semipilosum Fresen. var. *semipilosum*; Gillett, F.T.E.A. (Leguminosae–Papilionoideae 2): 1027. 1971.

Terrestrial herb. Montane grassland. 2360 m.

Distribution: MP, MR—*Gereau 3748*

Trifolium simense Fresen.; Gillett, F.T.E.A. (Leguminosae–Papilionoideae 2): 1022. 1971.

Terrestrial herb. Montane grassland. 2250–2640 m.

Distribution: KP, MP, RM—*Suleiman 56*

Trifolium tembense Fresen.; Gillett, F.T.E.A. (Leguminosae–Papilionoideae 2): 1033. 1971.

Terrestrial herb. Montane grassland. 2400–2570 m.

Distribution: KP—*Gereau 3105*

Trifolium wentzelianum Harms; Gillett, F.T.E.A. (Leguminosae–Papilionoideae 2): 1026. 1971.

Terrestrial herb. Montane grassland. 2250–2590 m.

Distribution: MP, MR—*Bullock 3978* (type: K), *Eggeling 6123*

Dietes iridioides (L.) Sweet ex Klatt; Goldblatt, F.T.E.A. (Iridaceae): 11. 1996.

Terrestrial herb. Riverine forest. 980–2340 m.

Distribution: MP, RM, SL—*Gereau 5170*

Gladiolus atropurpureus Baker; Goldblatt, F.T.E.A. (Iridaceae): 58. 1996.

Terrestrial herb. Woodland, montane grassland. 1400–2600 m.

Distribution: KP, MP, MR—*Gereau 3757*

Gladiolus curtifolius Marais; Goldblatt, F.T.E.A. (Iridaceae): 60. 1996.

Terrestrial herb. Woodland, montane grassland.

Distribution: MP, SL—*Goetze 1448* (type: B), *Stolz 2330*

Gladiolus dalenii Van Geel subsp. *andongensis* (Baker) Goldblatt; Goldblatt, F.T.E.A. (Iridaceae): 71. 1996.

Terrestrial herb. Woodland, riverine forest, heathland, montane grassland. 1200–2770 m.

Distribution: MP, RM, SL—*Gereau 7054*

Gladiolus dalenii Van Geel subsp. *dalenii*; Goldblatt, F.T.E.A. (Iridaceae): 70. 1996.

Terrestrial herb. Montane grassland. 2400–2580 m.

Distribution: KP, MP—*P. Lovett 15*

Gladiolus gregarius Welw. ex Baker; Goldblatt, F.T.E.A. (Iridaceae): 62. 1996.

Terrestrial herb. Miombo woodland. 1200–1530 m.

Distribution: MP, SL—*Gereau 4380*

Gladiolus laxiflorus Baker; Goldblatt, F.T.E.A. (Iridaceae): 55. 1996.

Terrestrial herb. Montane grassland. 2600 m.

Distribution: KP—*Leedal 4849*

Gladiolus sulculatus Goldblatt; Manning & Goldblatt, 2009: 45.

Terrestrial herb. Montane grassland. 2500–2600 m.

Distribution: KP, MR—*J. Lovett 806* (type: DSM, K, MO), *J. Lovett 1870*

Hesperantha cf. *petitiana* (A. Rich.) Baker; Goldblatt, F.T.E.A. (Iridaceae): 33. 1996.

Terrestrial herb. Rocky montane grassland. 2320 m.

Distribution: SL—*Gereau 3980*

Lapeirousia erythrantha (Klotzsch ex Klatt) Baker; Goldblatt, F.T.E.A. (Iridaceae): 23. 1996.

Terrestrial herb. Riverine forest edge. 1850 m.

Distribution: SL—*Gereau 4038*

Moraea callista Goldblatt; Goldblatt, F.T.E.A. (Iridaceae): 15. 1996.

Terrestrial herb. Montane grassland. 2320–2740 m.

Distribution: KP, SL—*Richards 7562* (type: K), *Gereau 3981*

Moraea schimperi (Hochst.) Pic. Serm.; Goldblatt, F.T.E.A. (Iridaceae): 18. 1996.

Terrestrial herb. Woodland, montane grassland. 1800–2700 m.

Distribution: KP, MP, NL, SL—*Gereau 4080*

Moraea tanzanica Goldblatt; Goldblatt, F.T.E.A. (Iridaceae): 22. 1996.

Terrestrial herb. Montane grassland. 2320–2700 m.

Distribution: KP, NL, RM, SL—*Stolz 2412* (type: K), *Suleiman 45*

Moraea thomsonii Baker; Goldblatt, F.T.E.A. (Iridaceae): 17. 1996.

Terrestrial herb. Montane grassland. 2440 m.

Distribution: KP—*Leedal 5521*

Moraea ventricosa Baker; Goldblatt, F.T.E.A. (Iridaceae): 21. 1996.

Terrestrial herb. Woodland. 1250–1300 m.

Distribution: MP—*J. Lovett 4621*

Moraea verdickii De Wild.; Goldblatt, F.T.E.A. (Iridaceae): 21. 1996.

Terrestrial herb. Montane grassland. 2600 m.

Distribution: KP—*Bidgood 137*

Romulea camerooniana Baker; Goldblatt, F.T.E.A. (Iridaceae): 36. 1996.

Terrestrial herb. Montane grassland. 2320–2740 m.

Distribution: KP, SL—*Gereau 3978*

Savannosiphon euryphyllus (Harms) Goldblatt & Marais; Goldblatt, F.T.E.A. (Iridaceae): 27. 1996.

Terrestrial herb. Roadsides in coffee plantation. 1310–1400 m.

Distribution: MP—*Gereau 7006*

JUNCACEAE

2 genera and 2 species

Juncus dregeanus Kunth subsp. *bachitii* (Hochst. ex Steud.) Hedb.; Haines & Lye, 1983: 32.

Terrestrial herb. Stream banks in montane grassland. 2130 m.

Distribution: MR, RM—*Wingfield 282*

Luzula abyssinica Parl.; Haines & Lye, 1983: 35.

Terrestrial herb. Montane grassland. 2700–2900 m.

Distribution: KP, MR, RM—*Pócs 6510/A*

LAMIACEAE

21 genera and 57 species

Achyrospermum carvalhoi Gürke var. *carvalhoi*; Polhill, F.T.E.A. (Lamiaceae): 34. 2009.

Shrub. Montane forest, riverine forest. 890–1710 m.

Distribution: RM, SL—*Mwasumbi 16287*

Achyrospermum cryptanthum Baker; Polhill, F.T.E.A. (Lamiaceae): 35. 2009.

Shrub. Montane forest. 2040–2400 m.

Distribution: MR, RM, SL—*Gereau 4703*

Aeollanthus subacaulis (Baker) Hua & Briq. var. *linearis* (Burkill) Ryding; Ryding, F.T.E.A. (Lamiaceae): 371. 2009.

Terrestrial herb. Montane grassland, riverine forest, miombo woodland. 1200–2488 m.

Distribution: MP, SL—*Suleiman 177*

Aeollanthus subacaulis (Baker) Hua & Briq. var. *subacaulis*; Ryding, F.T.E.A. (Lamiaceae): 371. 2009.

Terrestrial herb. Woodland. 1250 m.

Distribution: MP—*J. Lovett 4326*

Ajuga integrifolia Buch.-Ham. ex D. Don; Lukhoba, F.T.E.A. (Lamiaceae): 15. 2009.

Terrestrial herb. Montane grassland. 2570 m.

Distribution: KP—*Gereau 3106*

Clinopodium myrianthum (Baker) Ryding; Ryding, F.T.E.A. (Lamiaceae): 120. 2009.

Shrub. Montane forest edges. 2300 m.

Distribution: MR—*Gereau 4591*

Clinopodium simense (Benth.) Kuntze; Ryding, F.T.E.A. (Lamiaceae): 123. 2009.

Terrestrial herb. Thicket near waterfall in montane grassland. 2600–2650 m.

Distribution: KP—*Gereau 3190*

Clinopodium uhligii (Gürke) Ryding var. *obtusifolium* (Avetta) Ryding; Ryding, F.T.E.A. (Lamiaceae): 123. 2009.

Terrestrial herb. Montane grassland and forest edges. 2000–2700 m.

Distribution: KP, MP, NL, RM—*Gereau 4647*

Endostemon villosus (Briq.) M. Ashby; Paton, F.T.E.A. (Lamiaceae): 201. 2009.

Shrub. Miombo woodland. 1450 m.

Distribution: MP—*J. Lovett 4919*

Haumaniastrum caeruleum (Oliv.) P. A. Duvign. & Plancke; Paton, F.T.E.A. (Lamiaceae): 213. 2009.

Terrestrial herb. Woodland. 1200 m.

Distribution: MP—*Kayombo 2427*

Haumaniastrum uniflorum A. J. Paton; Paton, F.T.E.A. (Lamiaceae): 213. 2009.

Terrestrial herb. Grassland.

Distribution: MR, NL—*Davies 818*

Haumaniastrum villosum (Benth.) A. J. Paton; Paton, F.T.E.A. (Lamiaceae): 220. 2009.

Terrestrial herb. Miombo woodland. 1400 m.

Distribution: MP—*P. Lovett 101*

Hoslundia opposita Vahl; Paton, F.T.E.A. (Lamiaceae): 193. 2009.

Shrub. Riverine forest, miombo woodland. 1160–1500 m.

Distribution: MP, RM, SL—*Gereau 3938*

Leonotis decadonta Gürke var. *decadonta*; Iwarsson & Harvey, F.T.E.A. (Lamiaceae): 100. 2009.

Shrub.

Distribution: SL—*Iwarsson 1216*

Leonotis decadonta Gürke var. *porotoensis* Iwarsson & Y. B. Harv.; Iwarsson & Harvey, F.T.E.A. (Lamiaceae): 100. 2009.

Shrub. Montane grassland and forest margins. 2300–2650 m.

Distribution: MR, NL—*Iwarsson 1160* (type: BR, DSM, EA, K, NHT, S, UPS), *Richards 9727*

Leonotis goetzei Gürke; Iwarsson & Harvey, F.T.E.A. (Lamiaceae): 102. 2009.

Shrub. Montane grassland and shrubland, roadsides. 1500–2800 m.

Distribution: MR, NL—*Iwarsson 1193*

Leonotis grandis Iwarsson & Y. B. Harv.; Iwarsson & Harvey, F.T.E.A. (Lamiaceae): 97. 2009.

Shrub. Riverine forest. 1500 m.

Distribution: MP, NL—*J. Lovett 4695*

Leonotis myricifolia Iwarsson & Y. B. Harv.; Iwarsson & Harvey, F.T.E.A. (Lamiaceae): 101. 2009.

Shrub.

Distribution: KP, MP—*Myembe 40*

Leonotis ocymifolia (Burm. f.) Iwarsson var. *raineriana* (Vis.) Iwarsson; Iwarsson & Harvey, F.T.E.A. (Lamiaceae): 103. 2009.

Shrub or perennial herb. Montane forest edges, secondary bushland, woodland. 1300–2640 m.

Distribution: MP, MR, NL—*Gereau 4643*

Leucas deflexa Hook f. vel sp. aff.; Bramley, F.T.E.A. (Lamiaceae): 73. 2009.

Terrestrial herb. Abandoned cultivation. 1250 m.

Distribution: MP—*J. Lovett 4191*

Leucas tettensis Vatke; Bramley, F.T.E.A. (Lamiaceae): 91. 2009.

Terrestrial herb. Riverine forest. 1250–1350 m.

Distribution: MP—*J. Lovett 4662*

Micromeria imbricata (Forssk.) C. Chr. var. *imbricata*; Ryding, F.T.E.A. (Lamiaceae): 117. 2009.

Terrestrial herb or subshrub. Montane grassland. 1940–2400 m.

Distribution: MP, MR, SL—*Gereau 4680*

Ocimum decumbens Gürke; Paton, F.T.E.A. (Lamiaceae): 170. 2009.

Terrestrial herb. Montane grassland. 2230 m.

Distribution: SL—*Gereau 3589*

Ocimum fimbriatum Briq.; Paton, F.T.E.A. (Lamiaceae): 156. 2009.

Terrestrial herb. Woodland, rocky grassland. 1250–1810 m.

Distribution: MP, SL—*Harder 1497*

Ocimum gratissimum L. subsp. *gratissimum*; Paton, F.T.E.A. (Lamiaceae): 152. 2009.

Terrestrial herb. Woodland, riverine forest, disturbed montane forest. 1100–1790 m.

Distribution: MP, RM—*Abeid 2915*

Ocimum obovatum E. Mey. ex Benth. subsp. *cordatum* (A. J. Paton) A. J. Paton; Paton, F.T.E.A. (Lamiaceae): 169. 2009.

Terrestrial herb. Montane grassland, forest edges. 2650 m.

Distribution: MP—*Goyder 3847*

Orthosiphon schimperi Benth.; Paton, F.T.E.A. (Lamiaceae): 182. 2009.

Terrestrial herb. Woodland, riverine thickets. 1200–2000 m.

Distribution: MP, SL—*Gereau 4065*

Platostoma rotundifolium (Briq.) A. J. Paton; Paton, F.T.E.A. (Lamiaceae): 231. 2009.

Shrub or perennial herb. Montane grassland, woodland, forest edges, roadsides in cultivation. 1475–2240 m.

Distribution: MP, MR, NL, RM, SL—*Kayombo 2403*

Plectranthus adenophorus Gürke; Paton, F.T.E.A. (Lamiaceae): 280. 2009.

Terrestrial herb. Montane grassland, mountain ridges. 2100–2850 m.

Distribution: KP, MP, NL—*P. Lovett 12*

Plectranthus alboviolaceus Gürke; Paton, F.T.E.A. (Lamiaceae): 246. 2009.

Shrub. Montane forest. 2150 m.

Distribution: MR—*Gereau 4618*

Plectranthus autranii (Briq.) A. J. Paton; Paton, F.T.E.A. (Lamiaceae): 326. 2009.

Shrub. Montane forest.

Distribution: MR—*Goyder 3861*

Plectranthus daviesii (E. A. Bruce) B. Mathew; Paton, F.T.E.A. (Lamiaceae): 312. 2009.

Terrestrial herb.

Distribution: RM—*Davies R4* (type: K, SRGH), *Mgaza 468*

Plectranthus defoliatus Hochst ex Benth.; Paton, F.T.E.A. (Lamiaceae): 297. 2009.

Terrestrial herb. Montane grassland. 2300 m.

Distribution: MP, MR—*Iwarsson 1083*

Plectranthus densus N. E. Br.; Paton, F.T.E.A.

(Lamiaceae): 292. 2009.

Terrestrial herb. Montane grassland. 2150 m.

Distribution: SL—*Gereau 5138*

Plectranthus djalonensis (A. Chev.) A. J. Paton; Paton, F.T.E.A. (Lamiaceae): 286. 2009.

Terrestrial herb. Montane forest, miombo woodland. 1300–1660 m.

Distribution: MP, RM—*Kayombo 1008*

Plectranthus gracilis Suess.; Paton, F.T.E.A. (Lamiaceae): 269. 2009.

Terrestrial herb.

Distribution: RM—*Abdullah 1349*

Plectranthus ignotus A. J. Paton; Paton, F.T.E.A. (Lamiaceae): 278. 2009.

Terrestrial herb. Mountain slopes. 2080–2140 m.

Distribution: MP, SL—*Suleiman 87*

Plectranthus laxiflorus Benth.; Paton, F.T.E.A. (Lamiaceae): 250. 2009.

Terrestrial herb. Montane forest. 2280–2400 m.

Distribution: RM—*Mwasumbi 16217*

Plectranthus megacalyx A. J. Paton; Paton, F.T.E.A. (Lamiaceae): 316. 2009.

Terrestrial herb. Montane grassland.

Distribution: MR—*Geilinger 2884* (type: K), *Leedal 3835*

Plectranthus parvicalyx A. J. Paton; Paton, F.T.E.A. (Lamiaceae): 316. 2009.

Terrestrial herb. Montane grassland. 2400–2600 m.

Distribution: KP, MP—*J. Lovett 802*

Plectranthus pubescens Baker; Paton, F.T.E.A. (Lamiaceae): 272. 2009.

Terrestrial herb. Woodland. 1300–1600 m.

Distribution: MP, SL—*J. Lovett 4285*

Plectranthus punctatus (L. f.) L'Hér. subsp. *edulis* (Vatke) J. Paton; Paton, F.T.E.A. (Lamiaceae): 350. 2009.

Terrestrial herb. Montane grassland. 2360 m.

Distribution: MR, RM—*Gereau 3755*

Plectranthus punctatus (L. f.) L'Hér. subsp. *punctatus*; Paton, F.T.E.A. (Lamiaceae): 349. 2009.

Terrestrial herb.

Distribution: MR—*Goyder 3860*

Plectranthus rungwensis A. J. Paton; Paton, F.T.E.A. (Lamiaceae): 252. 2009.

Shrub. Montane forest. 2100–2350 m.

Distribution: MR, RM—*Mwasumbi 16222* (type: DSM, K, MO), *Richards 6514*

Plectranthus schizophyllus Baker; Paton, F.T.E.A. (Lamiaceae): 333. 2009.

Terrestrial herb. Montane grassland, miombo woodland. 1500–2700 m.

Distribution: KP, NL, SL—*Gereau 4418*

Plectranthus stolzii Gilli; Paton, F.T.E.A. (Lamiaceae): 254. 2009.

Terrestrial herb. Montane forest. 2040 m.

Distribution: MP, SL—*Gereau 4731*

Plectranthus viphyensis Brummitt & Seyani subsp. *viphyensis*; Paton, F.T.E.A. (Lamiaceae): 276. 2009.

Terrestrial herb. Woodland, montane grassland. 1400–1810 m.

Distribution: MP, SL—*Harder 1496*

Pycnostachys meyeri Gürke; Bramley, F.T.E.A. (Lamiaceae): 412. 2009.

Shrub or woody-based herb. Montane forest and brushland. 1610–1700 m.

Distribution: RM—*Mwasumbi 16338*

Pycnostachys ruandensis De Wild.; Bramley, F.T.E.A. (Lamiaceae): 389. 2009.

Terrestrial herb. Cultivated fields. 2280 m.

Distribution: MR—*Mwasumbi 16161*

Pycnostachys schliebenii Mildbr.; Bramley, F.T.E.A. (Lamiaceae): 410. 2009.

Shrub. Montane forest. 2000 m.

Distribution: RM—*Mwasumbi 16232*

Pycnostachys sphaerocephala Baker; Bramley, F.T.E.A. (Lamiaceae): 402. 2009.

Shrub. Woodland. 1400–1500 m.

Distribution: MP—*J. Lovett 3413*

Pycnostachys aff. *urticifolia* Hook.; Bramley, F.T.E.A. (Lamiaceae): 411. 2009.

Shrub. Montane forest edges. 2200 m.

Distribution: MR—*Gereau 4537*

Salvia nilotica Juss. ex Jacq.; Polhill, F.T.E.A. (Lamiaceae): 110. 2009.

Terrestrial herb. Montane grassland. 2640 m.

Distribution: KP, MP, NL—*Gereau 4638*

Scutellaria schweinfurthii Briq. subsp. *paucifolia* (Baker) A. J. Paton; Paton, F.T.E.A. (Lamiaceae): 28. 2009.

Terrestrial herb. Woodland. 1400–2250 m.

Distribution: MP—*J. Lovett 4862*

Stachys aculeolata Hook. f.; Harvey, F.T.E.A. (Lamiaceae): 42. 2009.

Herbaceous climber. Montane forest edges. 2200–2300 m.

Distribution: MR, SL—*Gereau 3532*

Stachys glandulibracteata Y. B. Harv.; Harvey, F.T.E.A. (Lamiaceae): 45. 2009.

Terrestrial herb. Streamsides in montane grassland. 1950–2100 m.

Distribution: MP, MR, RM—*Richards 17016* (type: EA, K), *Napper 1171*

Stachys pseudonigricans Gürke; Harvey, F.T.E.A. (Lamiaceae): 48. 2009.

Terrestrial herb. Woodland. 1300–2000 m.

Distribution: MP, SL—*Gereau 4064*

Tetradenia tanganyikae Phillipson; Phillipson, F.T.E.A. (Lamiaceae): 359. 2009.

Shrub. Riverine forest, woodland. 1150–1500 m.

Distribution: MP—*Kayombo 2441*

Teucrium goetzei Gürke; Paton, F.T.E.A. (Lamiaceae): 14. 2009.

Terrestrial herb. Montane grassland. 1700–1940 m.

Distribution: SL—*Goetze 1188a* (type: BM), *Gereau 4652*

Tinnea aethiopica Kotschy ex Hook. f. subsp. *stolzii* (Robyns & Lebrun) Vollesen; Willis, F.T.E.A. (Lamiaceae): 20. 2009.

Shrub. Woodland, riverine forest. 1200–1910 m.

Distribution: MP, SL—*Gereau 4071*

LAURACEAE

2 genera and 3 species

Cassytha filiformis L.; Verdcourt, F.T.E.A. (Lauraceae): 15. 1996.

Parasitic herb. Lakeshore forest. 474 m.

Distribution: SL—*Gereau 3639*

Cassytha pondoensis Engl.; Verdcourt, F.T.E.A. (Lauraceae): 17. 1996.

Parasitic herb. Secondary shrubland. 1760 m.

Distribution: NL—*Gereau 4245*

Ocotea usambarensis Engl.; Verdcourt, F.T.E.A. (Lauraceae): 10. 1996.

Tree. Montane forest. 1620–2050 m.

Distribution: NL, RM—*Abeid 2822*

LEEACEAE

1 genus and 1 species

Leea guineensis G. Don; Verdcourt, F.T.E.A. (Vitaceae [Leeaceae]): 141. 1993.

Terrestrial herb. Riverine forest. 890 m.

Distribution: SL—*Kibure 1469*

LENTIBULARIACEAE

1 genus and 1 species

Utricularia welwitschii Oliv.; Taylor, F.T.E.A. (Lentibulariaceae): 10. 1973.

Terrestrial herb. Marshy montane grassland. 2550 m.

Distribution: KP—*Abeid 3118*

LILIACEAE

1 genus and 1 species

Lilium formosanum Wallace; Grimshaw, F.T.E.A. (Liliaceae): 3. 2005.

Herb. Montane forest. 2100 m.

Distribution: MR—*Mwasumbi 16308*

LINACEAE

1 genus and 3 species

Linum keniense T. C. E. Fr.; Smith, F.T.E.A. (Linaceae): 10. 1966.

Terrestrial herb. Montane grassland.

Distribution: RM—*Procter 1449*

Linum thunbergii Eckl. & Zeyh.; Smith, F.T.E.A. (Linaceae): 10. 1966.

Terrestrial herb. Montane grassland. 2570–2700 m.

Distribution: KP, MP, NL—*Gereau 4087*

Linum volkensii Engl.; Smith, F.T.E.A. (Linaceae):

8. 1966.

Terrestrial herb. Montane grassland. 1940 m.

Distribution: SL—*Gereau 4662*

LOGANIACEAE

2 genera and 4 species

Anthocleista grandiflora Gilg; Leeuwenberg, Fl. Zamb. 7(1): 329. 1983.

Tree. Montane forest. 1610–1680 m.

Distribution: RM—*Kibure 1398*

Strychnos innocua Delile; Leeuwenberg, Fl. Zamb. 7(1): 361. 1983.

Tree. Miombo woodland. 1200 m.

Distribution: MP—*J. Lovett 3883*

Strychnos mellodora S. Moore vel sp. aff.; Leeuwenberg, Fl. Zamb. 7(1): 366. 1983.

Tree. Montane forest. 1920 m.

Distribution: NL—*Bracebridge 59*

Strychnos spinosa Lam.; Leeuwenberg, Fl. Zamb. 7(1): 371. 1983.

Tree. Riverine and lakeshore thickets, woodland, abandoned cultivation. 474–1250 m.

Distribution: MP, SL—*Mwasumbi 16602*

LORANTHACEAE

7 genera and 14 species

Agelanthus bipartitus Balle ex Polhill & Wiens; Polhill & Wiens, F.T.E.A. (Loranthaceae): 59. 1999.

Parasitic shrub. Montane forest.

Distribution: RM, SL—*Stolz 2475* (type: BM, BR, FHO, K, MO, P), *Richards 17622*

Agelanthus fuellebornii (Engl.) Polhill & Wiens; Polhill & Wiens, F.T.E.A. (Loranthaceae): 61. 1999.

Parasitic shrub. Woodland, sandy lakeshores. 478–1700 m.

Distribution: MP, SL—*Gereau 6084*

Agelanthus nyasicus (Baker & Sprague) Polhill & Wiens; Polhill & Wiens, F.T.E.A. (Loranthaceae): 62. 1999.

Parasitic shrub. Riverine forest, coffee plantations. 1300–1390 m.

Distribution: MP—*Gereau 7039*

Agelanthus subulatus (Engl.) Polhill & Wiens; Polhill & Wiens, F.T.E.A. (Loranthaceae): 46. 1999.

Parasitic shrub. Montane forest. 1900 m.

Distribution: NL—*Gereau 4183*

Agelanthus zizyphifolius (Engl.) Polhill & Wiens subsp. *vittatus* (Engl.) Polhill & Wiens; Polhill & Wiens, F.T.E.A. (Loranthaceae): 51. 1999.

Parasitic shrub. Montane forest edges. 1980 m.

Distribution: SL—*Gereau 3623*

Englerina cordata Polhill & Wiens; Polhill & Wiens, F.T.E.A. (Loranthaceae): 32. 1999.

Parasitic shrub. Woodland. 2110 m.

Distribution: SL—*Gereau 4018*

Englerina inaequilatera (Engl.) Gilli; Polhill & Wiens, F.T.E.A. (Loranthaceae): 31. 1999.

Parasitic shrub. Montane forest, forest edges. 1850–2640 m.

Distribution: KP, MP, MR, NL, RM, SL—*Gereau 4644*

Erianthemum taborense (Engl.) Tiegh.; Polhill & Wiens, F.T.E.A. (Loranthaceae): 98. 1999.

Parasitic shrub. Woodland. 1200–1475 m.

Distribution: MP—*Kayombo 2437*

Globimetula anguliflora (Engl.) Danser; Polhill & Wiens, F.T.E.A. (Loranthaceae): 79. 1999.

Parasitic shrub. Miombo woodland. 1500 m.

Distribution: MP—*Gereau 6985*

Oedina pendens (Engl. & K. Krause) Polhill & Wiens; Polhill & Wiens, F.T.E.A. (Loranthaceae): 86. 1999.

Parasitic shrub. Montane forest. 1910 m.

Distribution: NL, RM—*Stolz 1103* (type: B, K, WAG), *Gereau 4213*

Phragmanthera cornetii (Dewèvre) Polhill & Wiens; Polhill & Wiens, F.T.E.A. (Loranthaceae): 108. 1999.

Parasitic shrub. Woodland. 1400 m.

Distribution: MP—*J. Lovett 3340*

Phragmanthera dschallensis (Engl.) M. G. Gilbert; Polhill & Wiens, F.T.E.A. (Loranthaceae): 110. 1999.

Parasitic shrub. Woodland. 1200–1600 m.

Distribution: MP—*Kayombo 2396*

Phragmanthera usuiensis (Oliv.) M. G. Gilbert; Polhill & Wiens, F.T.E.A. (Loranthaceae): 105.

1999.

Parasitic shrub. Miombo woodland, montane forest. 1780–1890 m.

Distribution: RM, SL—*Gereau 5107*

Vanwykia rubella Polhill & Wiens; Polhill & Wiens, F.T.E.A. (Loranthaceae): 85. 1999.

Parasitic shrub. Woodland. 1300–1400 m.

Distribution: MP—*Kayombo 1041*

MALVACEAE

5 genera and 9 species

Hibiscus debeerstii De Wild. & T. Durand; Mwachala, F.T.E.A. (Malvaceae): 53. 2009.

Shrub. Riverine forest and edges of miombo woodland. 1200–1500 m.

Distribution: MP, SL—*Kayombo 2366*

Hibiscus fuscus Garcke; Mwachala, F.T.E.A. (Malvaceae): 49. 2009.

Shrub. Montane forest. 1870 m.

Distribution: RM—*WCS SHCP 231*

Hibiscus ludwigii Eckl. & Zeyh.; Mwachala, F.T.E.A. (Malvaceae): 65. 2009.

Scrambling shrub. Montane forest, woodland. 1400–2100 m.

Distribution: MP, MR—*Mwasumbi 16318*

Hibiscus aff. *micranthus* L. f.; Mwachala, F.T.E.A. (Malvaceae): 47.

Shrub or perennial herb. Montane forest, *Eucalyptus* plantations, riverine forest. 1200–2100 m.

Distribution: MP, MR, RM—*Mwasumbi 16337*

Malvaviscus penduliflorus DC.; Turner & Mendenhall, 1993: 449.

Cultivated shrub. Garden in village. 1250 m.

Distribution: MP—*Kayombo 2337*

Pavonia kilimandscharica Gürke; Verdcourt, F.T.E.A. (Malvaceae): 14. 2009.

Shrub. Montane forest. 2080 m.

Distribution: MR, RM—*Mwasumbi 16187*

Pavonia urens Cav.; Verdcourt, F.T.E.A. (Malvaceae): 11. 2009.

Shrub. Montane forest, forest edges, riverine forest. 1400–2360 m.

Distribution: MP, MR, RM, SL—*Gereau 4707*

Sida rhombifolia L. var. *serratifolia* (R. Wilczek & Steyaert) Verdc.; Verdcourt, F.T.E.A. (Malvaceae): 158. 2009.

Shrub. Montane forest. 2080 m.

Distribution: MR—*Mwasumbi 16188*

Thespesia garckeana F. Hoffm.; Verdcourt, F.T.E.A. (Malvaceae): 104. 2009.

Tree or shrub. Riverine forest, woodland. 1100–1410 m.

Distribution: MP, SL—*Kayombo 960*

MELASTOMATACEAE

2 genera and 7 species

Antherotoma naudinii Hook. f.; Wickens, F.T.E.A. (Melastomataceae): 9. 1975.

Terrestrial herb. Woodland, grassland, riverine forest, abandoned cultivation. 1250–1940 m.

Distribution: MP, SL—*Gereau 4668*

Dissotis cryptantha Baker; Wickens, F.T.E.A. (Melastomataceae): 56. 1975.

Shrub.

Distribution: RM—*Goetze 1138*

Dissotis melleri Hook. f. var. *greenwayi* (A. Fern. & R. Fern.) & R. Fern.; Wickens, F.T.E.A. (Melastomataceae): 46. 1975.

Shrub. Montane forest edges, bamboo thicket. 1810–2350 m.

Distribution: MR, NL, RM, SL—*Gereau 3971*

Dissotis pachytricha Gilg ex R. E. Fr. var. *orientalis* A. Fern. & R. Fern.; Wickens, F.T.E.A. (Melastomataceae): 54. 1975.

Shrub. Thickets in montane forest. 1840 m.

Distribution: SL—*Gereau 3541*

Dissotis princeps (Kunth) Triana var. *princeps*; Wickens, F.T.E.A. (Melastomataceae): 51. 1975.

Shrub. Open woodland. 1650–1680 m.

Distribution: SL—*Gereau 5117*

Dissotis senegambiensis (Guill. & Perr.) Triana var. *alpestris* (Taub.) A. Fern. & R. Fern.; Wickens, F.T.E.A. (Melastomataceae): 50. 1975.

Terrestrial herb. Montane grassland. 1940 m.

Distribution: SL—*Gereau 4684*

Dissotis trothae Gilg; Wickens, F.T.E.A. (Melastomataceae): 55. 1975.

Shrub. Stream banks in montane grassland. 2080–2140 m.

Distribution: SL—*Suleiman 119*

MELIACEAE

4 genera and 9 species

Lepidotrichilia volkensii (Gürke) J.-F. Leroy ex Styles & F. White; Styles & White, F.T.E.A. (Meliaceae): 37. 1991.

Tree or shrub. Riverine forest, montane forest. 1730–2400 m.

Distribution: MP, MR, NL, RM—*Abeid 2900*

Toona ciliata M. Roem.; Styles & White, F.T.E.A. (Meliaceae): 45. 1991.

Tree. Lakeshore forest and thickets; naturalized. 474–490 m.

Distribution: SL—*Mwasumbi 16559*

Trichilia dregeana Sond.; Styles & White, F.T.E.A. (Meliaceae): 34. 1991.

Tree. Riverine forest. 1350 m.

Distribution: MP—*J. Lovett 4757*

Trichilia emetica Vahl; Styles & White, F.T.E.A. (Meliaceae): 34. 1991.

Tree. Lakeshore forest, riverine forest, coffee plantations. 474–1300 m.

Distribution: MP, SL—*Gereau 3671*

Trichilia aff. *prieuriana* A. Juss.; Styles & White, F.T.E.A. (Meliaceae): 30. 1991.

Tree. Riverine forest. 1275–1400 m.

Distribution: MP—*Kayombo 2378*

Turraea floribunda Hochst.; Styles & White, F.T.E.A. (Meliaceae): 21. 1991.

Tree. Montane forest.

Distribution: RM—*Ndangalasi 977*

Turraea holstii Gürke; Styles & White, F.T.E.A. (Meliaceae): 13. 1991.

Shrub. Woodland, riverine forest. 1250–1600 m.

Distribution: MP—*Kayombo 2352*

Turraea robusta Gürke; Styles & White, F.T.E.A. (Meliaceae): 8. 1991.

Tree. Riverine forest. 1500–1780 m.

Distribution: RM, SL—*Kibure 1427*

Turraea stolzii Harms; Styles & White, F.T.E.A. (Meliaceae): 13. 1991.

Shrub. Riverine brushland, montane forest edges. 1500–1660 m.

Distribution: RM—*Mwasumbi 16326*

MELIANTHACEAE

1 genus and 1 species

Bersama abyssinica Fresen. subsp. *abyssinica*; Verdcourt, F.T.E.A. (Melianthaceae): 4. 1958.

Tree. Riverine forest, montane forest, scree slopes. 1650–2260 m.

Distribution: MP, MR, SL—*Suleiman 138*

Bersama abyssinica Fresen. subsp. *paulliniodes* (Planch.) Verdc.; Verdcourt, F.T.E.A. (Melianthaceae): 5. 1958.

Tree. Riverine forest, montane forest. 1770–2040 m.

Distribution: NL, SL—*Gereau 4033*

MENISPERMACEAE

3 genera and 3 species

Stephania abyssinica (Quart.-Dill. & A. Rich.) Walp. var. *abyssinica*; Troupin, F.T.E.A. (Menispermaceae): 22. 1956.

Herbaceous climber. Tree clumps in montane grassland. 2360 m.

Distribution: MR, RM—*Gereau 3718*

Tiliacora funifera (Miers) Oliv.; Troupin, F.T.E.A. (Menispermaceae): 9. 1956.

Liana. Disturbed montane forest. 1790–1910 m.

Distribution: RM—*Abeid 2910*

Tinospora caffra (Miers) Troupin; Troupin, F.T.E.A. (Menispermaceae): 19. 1956.

Herbaceous climber. Lakeshore forest. 474 m.

Distribution: SL—*Gereau 3670*

MIMOSACEAE

7 genera and 18 species

Acacia amythethophylla Steud. ex A. Rich.; Asfaw & Thulin, 1989: 86.

Tree. Woodland, abandoned cultivation. 1100–1530 m.

Distribution: MP, SL—*Kayombo 2385*

Acacia erubescens Welw. ex Oliv.; Brenan, F.T.E.A. (Leguminosae–Mimosoideae): 88. 1959.

Shrub or tree. Woodland. 1200 m.

Distribution: MP—*J. Lovett 3857*

Acacia polyacantha Willd. subsp. *campylacantha* (Hochst. ex A. Rich.) Brenan; Brenan, F.T.E.A. (Leguminosae–Mimosoideae): 88. 1959.

Tree. Woodland, riverine forest, lakeshore forest, cultivated fields. 480–1400 m.

Distribution: MP, SL—*Gereau 3660*

Acacia pseudofistula Harms vel sp. aff.; Brenan, F.T.E.A. (Leguminosae–Mimosoideae): 122. 1959.

Shrub. Woodland. 1200–1300 m.

Distribution: MP—*P. Lovett 379*

Acacia schweinfurthii Brenan & Exell var. *schweinfurthii*; Brenan, F.T.E.A. (Leguminosae–Mimosoideae): 99. 1959.

Liana. Riverine forest. 1350 m.

Distribution: MP—*J. Lovett 3604*

Acacia sieberiana DC. var. *woodii* (Burtt Davy) Keay & Brenan; Asfaw & Thulin, 1989: 90.

Tree. Woodland, riverine forest. 1250–1600 m.

Distribution: MP—*Gereau 6092*

Albizia amara (Roxb.) Boivin; Brenan, F.T.E.A. (Leguminosae–Mimosoideae): 151. 1959.

Tree. Montane forest. 1510 m.

Distribution: RM—*WCS SHCP 320*

Albizia antunesiana Harms; Brenan, F.T.E.A. (Leguminosae–Mimosoideae): 148. 1959.

Tree. Woodland. 1280–1400 m.

Distribution: MP—*Kayombo 2367*

Albizia gummifera (J. F. Gmel.) C. A. Sm. var. *gummifera*; Brenan, F.T.E.A. (Leguminosae–Mimosoideae): 157. 1959.

Tree. Riverine forest. 1300–1500 m.

Distribution: MP, RM—*J. Lovett 4788*

Albizia harveyi E. Fourn.; Brenan, F.T.E.A. (Leguminosae–Mimosoideae): 149. 1959.

Tree. Woodland. 1250 m.

Distribution: MP—*J. Lovett 4368*

Albizia lebbeck (L.) Benth.; Brenen, F.T.E.A. (Leguminosae–Mimosoideae): 147. 1959.

Tree.

Distribution: NL—*Davies 645*

Albizia schimperiana Oliv. var. *schimperiana*; Brenan, F.T.E.A. (Leguminosae–Mimosoideae): 154. 1959.

Tree. Riverine forest. 1400 m.

Distribution: MP, NL—*J. Lovett 4686*

Albizia versicolor Welw. ex Oliv.; Brenan, F.T.E.A. (Leguminosae–Mimosoideae): 146. 1959.

Tree. Woodland. 1300–1400 m.

Distribution: MP—*Gereau 6100*

Dichrostachys cinerea (L.) Wight & Arn. subsp. *nyassana* (Taub.) Brenan; Brenan & Brummitt, Fl. Zamb. 3(1): 40. 1970.

Tree or shrub. Woodland. 1200–1310 m.

Distribution: MP—*Gereau 6103*

Entada abyssinica Steud. ex A. Rich.; Brenan, F.T.E.A. (Leguminosae–Mimosoideae): 13. 1959.

Tree. Roadsides, thickets, miombo woodland, cultivation. 1250–1475 m.

Distribution: MP—*Kayombo 2384*

Faidherbia albida (Delile) A. Chev.; Barnes & Fagg, 2003: 3.

Tree. Coffee plantations. 1250–1400 m.

Distribution: MP—*Kayombo 957*

Mimosa pigra L.; Brenan, F.T.E.A. (Leguminosae–Mimosoideae): 43. 1959.

Lianescent shrub. Lakeshore forest. 480 m.

Distribution: SL—*Gereau 3658*

Parkia filicoidea Welw. ex Oliv.; Brenan, F.T.E.A. (Leguminosae–Mimosoideae): 7. 1959.

Tree. Riverine forest. 1160 m.

Distribution: SL—*Gereau 3703*

MOLLUGINACEAE

1 genus and 1 species

Glinus lotoides L. var. *virens* Fenzl; Gonçalves, Fl. Zamb. 4: 538. 1978.

Terrestrial herb. Woodland. 1250 m.

Distribution: MP—*J. Lovett 3775*

MONIMIACEAE

1 genus and 1 species

Xymalos monospora (Harv.) Baill. ex Warb.; Verdcourt, F.T.E.A. (Monimiaceae): 1. 1968.

Shrub or tree. Montane forest, riverine forest. 1780–2300 m.

Distribution: MP, MR, NL, RM, SL—*Gereau 4745*

MORACEAE

3 genera and 13 species

Dorstenia benguellensis Welw.; Hijman, F.T.E.A. (Moraceae): 38. 1989.

Terrestrial herb. Miombo woodland, riverine forest edges. 1300–1850 m.

Distribution: MP, SL—*Gereau 3901*

Dorstenia cuspidata Hochst. ex A. Rich.; Hijman, F.T.E.A. (Moraceae): 40. 1989.

Terrestrial herb. Miombo woodland, riverine forest. 1160–1530 m.

Distribution: SL—*Gereau 4372*

Dorstenia psilurus Welw.; Hijman, F.T.E.A. (Moraceae): 31. 1989.

Terrestrial herb. Riverine forest. 1320 m.

Distribution: MP—*Gereau 6121*

Dorstenia schliebenii Mildbr.; Hijman, F.T.E.A. (Moraceae): 28. 1989.

Terrestrial herb. Montane forest. 1730 m.

Distribution: NL—*Gereau 4147*

Ficus exasperata Vahl; Berg, F.T.E.A. (Moraceae): 52. 1989.

Tree. Riverine forest, woodland, coffee plantations, abandoned cultivation. 1250–1400 m.

Distribution: MP—*Gereau 6098*

Ficus ingens (Miq.) Miq.; Berg, F.T.E.A. (Moraceae): 60. 1989.

Tree. Lakeshore bushland. 474–480 m.

Distribution: SL—*Kibure 1490*

Ficus natalensis Hochst.; Berg, F.T.E.A. (Moraceae): 71. 1989.

Tree. Riverine forest. 1290 m.

Distribution: MP—*Gereau 6109*

Ficus ottoniifolia (Miq.) Miq. subsp. *ulugurensis* (Warb. ex Mildbr. & Burret) C. C. Berg; Berg, F.T.E.A. (Moraceae): 77. 1989.

Tree. Riverine forest, montane forest. 1250–1840 m.

Distribution: MP, RM—*Kayombo 2480*

Ficus sur Forssk.; Berg, F.T.E.A. (Moraceae): 56. 1989.

Tree. Riverine forest. 1160–1850 m.

Distribution: MP, SL—*Gereau 3694*

Ficus sycomorus L.; Berg, F.T.E.A. (Moraceae): 54. 1989.

Tree. Woodland, riverine forest, coffee plantations. 1250–1370 m.

Distribution: MP—*Gereau 6106*

Ficus thonningii Blume; Berg, F.T.E.A. (Moraceae): 73. 1989.

Tree. Riverine forest. 1500 m.

Distribution: SL—*Gereau 4795*

Ficus vallis-choudae Delile; Berg, F.T.E.A. (Moraceae): 58. 1989.

Tree. Riverine forest. 580–1150 m.

Distribution: MP, SL—*Kibure 1448*

Trilepisium madagascariense DC.; Berg, F.T.E.A. (Moraceae): 17. 1989.

Tree. Riverine forest. 890–1680 m.

Distribution: MP, RM, SL—*Kayombo 1097*

MUSACEAE

1 genus and 1 species

Ensete ventricosum (Welw.) Cheesman; Lock, F.T.E.A. (Musaceae): 3. 1993.

Terrestrial herb. Woodland. 1750–1950 m.

Distribution: MP, RM—*Lovett 3431*

MYRICACEAE

1 genus and 2 species

Morella salicifolia (Hochst. ex A. Rich.) Verdc. & Polhill subsp. *kilimandscharica* (Engl.) Verdc. & Polhill; Polhill & Verdcourt, F.T.E.A. (Myricaceae): 9. 2000.

Tree. Montane forest, tree clumps in grassland. 1670–2380 m.

Distribution: KP, MP, MR, RM, SL—*Gereau 3736*

Morella serrata (Lam.) Killick; Polhill & Verdcourt, F.T.E.A. (Myricaceae): 4. 2000.

Tree. Riverine forest. 1500 m.

Distribution: SL—*Gereau 4782*

MYROTHAMNACEAE

1 genus and 1 species

Myrothamnus flabellifolius Welw.; Bywater, F.T.E.A. (Myrothamnaceae): 1. 1984.

Shrub. Rocky mountain slopes. 2320 m.

Distribution: SL—*Gereau 3993*

MYRSINACEAE

3 genera and 4 species

Embelia schimperi Vatke; Halliday, F.T.E.A. (Myrsinaceae): 13. 1984.

Liana. Montane forest, riverine forest. 1500–2430 m.

Distribution: MR, RM, SL—*Abeid 2814*

Maesa lanceolata Forssk.; Halliday, F.T.E.A. (Myrsinaceae): 3. 1984.

Shrub or tree. Montane forest, riverine forest, miombo woodland, bamboo thicket. 1500–2430 m.

Distribution: KP, MP, MR, NL, SL—*Suleiman 133*

Myrsine africana L.; Halliday, F.T.E.A. (Myrsinaceae): 6. 1984.

Shrub. Montane heathland and shrubland. 2620–2820 m.

Distribution: KP, RM—*Mwasumbi 16253*

Myrsine melanophloeos (L.) R. Br.; Lovett et al., 2006: 188.

Shrub or tree. Montane forest edges, miombo woodland. 620–2700 m.

Distribution: MR, NL, RM, SL—*Gereau 5122*

MYRTACEAE

1 genus and 3 species

Syzygium cordatum Hochst. subsp. *cordatum*; Verdcourt, F.T.E.A. (Myrtaceae): 74. 2001.

Tree or shrub. Riverine forest, secondary shrubland. 1530–1810 m.

Distribution: MP, NL, SL—*Gereau 4243*

Syzygium guineense (Willd.) DC. subsp. *afromontanum* F. White; Verdcourt, F.T.E.A. (Myrtaceae): 79. 2001.

Tree. Riverine forest, lower montane forest, woodland. 1400–1810 m.

Distribution: MP, NL, SL—*Harder 1502*

Syzygium guineense (Willd.) DC. subsp. *guineense*; Verdcourt, F.T.E.A. (Myrtaceae): 78. 2001.

Tree. Riverine forest, woodland, edges of cultivation. 474–1400 m.

Distribution: MP, SL—*Mwasumbi 16590*

Syzygium guineense (Willd.) DC. subsp. *macrocarpum* (Engl.) F. White; Verdcourt, F.T.E.A. (Myrtaceae): 79. 2001.

Tree. Riverine forest. 1300 m.

Distribution: MP—*J. Lovett 3594*

Syzygium masukuense (Baker) R. E. Fr. subsp. *masukuense*; Verdcourt, F.T.E.A. (Myrtaceae): 82. 2001.

Tree. Montane forest. 2010 m.

Distribution: RM—*WCS SHCP 14*

OCHNACEAE

1 genus and 7 species

Ochna afzelii R. Br. ex Oliv. subsp. *mechowiana* (O. Hoffm.) N. Robson; Verdcourt, F.T.E.A. (Ochnaceae): 34. 2005.

Shrub. Miombo woodland. 1780 m.

Distribution: SL—*Gereau 5108*

Ochna holstii Engl.; Verdcourt, F.T.E.A. (Ochnaceae): 23. 2005.

Tree. Riverine forest, montane forest. 1450–1990 m.

Distribution: MP, RM—*Kibure 1413*

Ochna oxyphylla N. Robson; Verdcourt, F.T.E.A. (Ochnaceae): 24. 2005.

Shrub or tree. Forest patches above treeline, montane forest edges. 2150–2320 m.

Distribution: MR, SL—*Gereau 3982*

Ochna polyneura Gilg vel sp. aff.; Verdcourt, F.T.E.A. (Ochnaceae): 25. 2005.

Suffrutex. Riverine forest. 1200–1300 m.

Distribution: MP—*J. Lovett 4906*

Ochna puberula N. Robson; Verdcourt, F.T.E.A. (Ochnaceae): 26. 2005.

Tree or shrub. Woodland. 1400–2000 m.

Distribution: MP, SL—*Gereau 4021*

Ochna richardsiae N. Robson; Verdcourt, F.T.E.A. (Ochnaceae): 31. 2005.

Suffrutex. Woodland. 1300–1600 m.

Distribution: MP—*J. Lovett 3612*

Ochna schweinfurthiana F. Hoffm.; Verdcourt, F.T.E.A. (Ochnaceae): 33. 2005.

Shrub or tree. Woodland. 1300–1400 m.
Distribution: MP—*J. Lovett 4810*

OLACACEAE

3 genera and 3 species

Olax dissitiflora Oliv.; Lucas, F.T.E.A. (Olacaceae): 9. 1968.
Shrub. Lakeshore bushland. 474–490 m.
Distribution: SL—*Mwasumbi 16571*

Strombosia scheffleri Engl.; Lucas, F.T.E.A. (Olacaceae): 11. 1968.
Tree. Riverine forest, montane forest. 910–1950 m.
Distribution: MP, RM, NL, SL—*Kayombo 1071*

Ximenia americana L.; Lucas, F.T.E.A. (Olacaceae): 3. 1968.
Tree. Miombo woodland. 1350 m.
Distribution: MP—*Kayombo 1091*

OLEACEAE

3 genera and 8 species

Jasminum abyssinicum Hochst. ex DC.; Turrill, F.T.E.A. (Oleaceae): 18. 1952.
Liana. Montane forest. 1580–1900 m.
Distribution: KP, RM—*Mwasumbi 16331*

Jasminum fluminense Vell.; Turrill, F.T.E.A. (Oleaceae): 19. 1952.
Liana. Disturbed montane forest.
Distribution: RM—*Ndangalasi 961*

Jasminum odoratissimum L. subsp. *goetzeanum* (Gilg) P. S. Green; Kupicha, Fl. Zamb. 7(1): 310. 1983.
Lianescent shrub. Montane forest and thickets, tree clumps in grassland. 1650–2360 m.
Distribution: MR, RM, SL—*Gereau 4716*

Jasminum schimperi Vatke; Kupicha, Fl. Zamb. 7(1): 318. 1983.
Liana. Wooded grassland. 1270 m.
Distribution: MP—*Gereau 3476*

Jasminum streptopus E. Mey.; Kupicha, Fl. Zamb. 7(1): 315. 1983.
Twining shrub. Woodland, abandoned cultivation, road banks. 1200–1400 m.
Distribution: MP—*Gereau 7007*

Olea capensis L. subsp. *macrocarpa* (C. H. Wright) I. Verd.; Lovett et al., 2006: 197.
Tree. Montane forest. 1840–2340 m.
Distribution: NL, RM—*Abeid 2993*

Schrebera alata (Hochst.) Welw.; Turrill, F.T.E.A. (Oleaceae): 4. 1952.
Tree. Riverine forest, montane thickets, woodland, abandoned cultivation. 1250–1850 m.
Distribution: RM, SL—*Gereau 4026*

Schrebera trichoclada Welw.; Turrill, F.T.E.A. (Oleaceae): 2. 1952.
Tree or shrub. Woodland. 1200–1400 m.
Distribution: MP—*J. Lovett 4337*

OLINIACEAE

1 genus and 1 species

Olinia rochetiana A. Juss.; Verdcourt, F.T.E.A. (Oliniaceae): 2. 1975.
Tree. Montane forest, riverine forest, montane heathland. 1700–2770 m.
Distribution: MP, NL, RM, SL—*Abeid 2811*

ONAGRACEAE

3 genera and 4 species

Epilobium capense Buchinger ex Hochst.; Raven, Fl. Zamb. 4: 343. 1978.
Terrestrial herb. Montane forest edges. 2360 m.
Distribution: MR—*Gereau 3729*

Epilobium salignum Hausskn.; Raven, Fl. Zamb. 4: 345. 1978.
Terrestrial herb. Montane grassland, roadsides. 2210–2550 m.
Distribution: KP, MR, RM—*Suleiman 43*

Ludwigia abyssinica A. Rich.; Raven, Fl. Zamb. 4: 340. 1978.
Terrestrial herb. Wet meadows, riverine forest. 580–1250 m.
Distribution: MP, SL—*Gereau 3804*

Oenothera glazioviana Micheli; Dietrich et al., 1997: 107.
Shrub. Secondary brushland. 1610 m.
Distribution: RM—*Mwasumbi 16336*

OPILIACEAE

1 genus and 1 species

Opilia amentacea Roxb.; Hiepko, 1989: 358.

Liana. Montane forest and thickets, lakeshore forest. 474–2230 m.

Distribution: MR, NL, SL—*Gereau 4214*

ORCHIDACEAE

28 genera and 97 species

Aerangis carnea J. L. Stewart; Stewart, F.T.E.A. (Orchidaceae 3): 558. 1989.

Epiphyte. Riverine forest. 1400–1830 m.

Distribution: MP, MR—*Cunningham s.n.* (type: K), *J. Lovett 4819*

Aerangis oligantha Schltr.; Stewart, F.T.E.A. (Orchidaceae 3): 565. 1989.

Epiphyte. Montane forest. 1800 m.

Distribution: MR, RM—*Stolz 2399* (type: K), *Pettersson 485*

Angraecum chamaeanthus Schltr.; Cribb, F.T.E.A. (Orchidaceae 3): 494. 1989.

Epiphyte. Montane forest.

Distribution: RM—*Richards 14306*

Angraecum stolzii Schltr.; Cribb, F.T.E.A. (Orchidaceae 3): 498. 1989.

Epiphyte. Montane forest. 2000–2500 m.

Distribution: MR, RM—*Stolz 2508* (type: B, K), *Cribb 11318*

Ansellia africana Lindl.; Cribb, F.T.E.A. (Orchidaceae 2): 402. 1984.

Epiphyte. Montane forest. 1675 m.

Distribution: RM—*WCS SHCP 254*

Bonatea stereophylla (Kraenzl.) Summerh.; Summerhayes, F.T.E.A. (Orchidaceae 1): 140. 1968.

Terrestrial herb. Montane grassland. 1500 m.

Distribution: RM—*Goetze 1339* (type: B [destroyed])

Brachycorythis pleistophylla Rchb. f.; Summerhayes, F.T.E.A. (Orchidaceae 1): 22. 1968.

Terrestrial herb. Miombo woodland, abandoned cultivation. 1780–1790 m.

Distribution: MP, SL—*Gereau 3879A*

Brownleea parviflora Harv. ex Lindl.; Summerhayes, F.T.E.A. (Orchidaceae 1): 179. 1968.

Terrestrial herb. Montane grassland, cypress plantation. 1800–2740 m.

Distribution: KP, MP, MR, RM—*J. Lovett 1875*

Bulbophyllum flectens P. J. Cribb & P. Taylor; Cribb, F.T.E.A. (Orchidaceae 2): 307. 1984.

Epiphyte. Montane forest.

Distribution: RM—*Leedal 1341*

Bulbophyllum longiflorum Thouars; Cribb, F.T.E.A. (Orchidaceae 2): 304. 1984.

Epiphyte. Forest.

Distribution: RM—*Cribb 10666*

Bulbophyllum mahonii Rolfe; Cribb, F.T.E.A. (Orchidaceae 2): 310. 1984.

Epiphyte. Montane forest.

Distribution: NL—*Cribb 10626*

Bulbophyllum cf. *mahonii* Rolfe; Cribb, F.T.E.A. (Orchidaceae 2): 310. 1984.

Epiphyte. Disturbed montane forest.

Distribution: RM—*Ndangalasi 1020*

Calanthe sylvatica (Thouars) Lindl.; Cribb, F.T.E.A. (Orchidaceae 2): 282. 1984.

Terrestrial herb. Montane forest. 1620–2600 m.

Distribution: MR, NL, RM—*Suleiman 76*

Corycium nigrescens Sond.; Cribb & Kurzweil, 1994: 555.

Terrestrial herb. Montane grassland.

Distribution: KP—*Leedal 4964*

Cynorkis anacamptoides Kraenzl.; Summerhayes, F.T.E.A. (Orchidaceae 1): 32. 1968.

Terrestrial herb. Marshes and stream banks in montane grassland. 2400–2600 m.

Distribution: KP, RM, SL—*Suleiman 49*

Cynorkis kaessneriana Kraenzl.; Summerhayes, F.T.E.A. (Orchidaceae 1): 33. 1968.

Terrestrial herb. Open volcanic ridges, *Protea* woodland. 1800–2550 m.

Distribution: RM—*Abeid 2838*

Cynorkis rungweensis Schltr.; Summerhayes, F.T.E.A. (Orchidaceae 1): 35. 1968.

Terrestrial herb. Montane grassland. 2000 m.

Distribution: RM—*Stolz 2532* (type: K)

Cynorkis symoensii Geerinck & Tournay; Geerinck, 1984: 165.

Terrestrial herb. Open volcanic ridges. 1800–2240 m.

Distribution: RM—*Bidgood 96*

Note: New to East Africa, from Malawi and Rwanda.

Diaphananthe meliantha P. J. Cribb; Cribb, F.T.E.A. (Orchidaceae 3): 540. 1989.

Epiphyte. Montane forest. 2070 m.

Distribution: NL—*Gereau 4141*

Disa aconitoides Sond. subsp. *concinna* (N. E. Br.) H. P. Linder; Linder, 1981: 91.

Terrestrial herb. Woodland. 1250–1400 m.

Distribution: MP—*J. Lovett 5037*

Disa aconitoides Sond. subsp. *goetzeana* (Kraenzl.) H. P. Linder; Linder, 1981: 92.

Terrestrial herb. Miombo woodland. 1790 m.

Distribution: SL—*Gereau 3883*

Disa equestris Rchb. f. vel sp. aff.; Linder, 1981: 99.

Terrestrial herb. Rocky mountain summit. 2800 m.

Distribution: KP—*Goldblatt 8206*

Disa erubescens Rendle; Summerhayes, F.T.E.A. (Orchidaceae 1): 157. 1968.

Terrestrial herb. Montane grassland. 2400–2700 m.

Distribution: KP, NL, SL—*Suleiman 172*

Disa fragrans Schltr. subsp. *fragrans*; Linder, 1999: 197.

Terrestrial herb. Montane grassland. 2740 m.

Distribution: KP, MP—*Wingfield 952*

Disa hamatopetala Rendle; Summerhayes, F.T.E.A. (Orchidaceae 1): 177. 1968.

Terrestrial herb. Montane grassland.

Distribution: KP—*Willan 176*

Disa longilabris Schltr.; Summerhayes, F.T.E.A. (Orchidaceae 1): 177. 1968.

Terrestrial herb. Montane grassland. 2550–2760 m.

Distribution: KP, RM, SL—*Goetze 1222, 1226* (syntypes: B [destroyed]), *Abeid 2837*

Disa ochrostachya Rchb. f.; Summerhayes, F.T.E.A. (Orchidaceae 1): 164. 1968.

Terrestrial herb. Montane grassland. 2440 m.

Distribution: SL—*Suleiman 183*

Disa robusta N. E. Br.; Summerhayes, F.T.E.A. (Orchidaceae 1): 168. 1968.

Terrestrial herb. Montane grassland. 2150–2700 m.

Distribution: KP, MR, RM—*Gereau 3781*

Disa rungweensis Schltr.; Summerhayes, F.T.E.A. (Orchidaceae 1): 175. 1968.

Terrestrial herb. Rocky mountain slopes. 2400–2960 m.

Distribution: RM—*Stolz 2437* (type: K), *Richards 14302*

Disa stolzii Schltr.; Summerhayes, F.T.E.A. (Orchidaceae 1): 156. 1968.

Terrestrial herb. Montane grassland, rocky slopes. 2600–2800 m.

Distribution: KP, RM—*Stolz 2506* (type: K), *Abeid 3092*

Disa walleri Rchb. f.; Summerhayes, F.T.E.A. (Orchidaceae 1): 169. 1968.

Terrestrial herb. Miombo woodland. 1450 m.

Distribution: MP—*Gereau 6982*

Disa walteri Schltr.; Summerhayes, F.T.E.A. (Orchidaceae 1): 176. 1968.

Terrestrial herb. Montane grassland. 2200–2800 m.

Distribution: MP, SL—*Goetze 925* (type: B [destroyed]), *Kerfoot 1632*

Disa welwitschii Rchb. f. vel sp. aff.; Summerhayes, F.T.E.A. (Orchidaceae 1): 160. 1968.

Terrestrial herb. Rocky montane grassland. 2720 m.

Distribution: KP—*Abeid 3055*

Disperis anthoceros Rchb. f.; Verdcourt, F.T.E.A. (Orchidaceae 1): 229. 1968.

Terrestrial herb. Montane forest, woodland. 1250–2140 m.

Distribution: MP, SL—*Suleiman 103*

Eulophia clavicornis Lindl. var. *nutans* (Sond.) A. V. Hall; Cribb, F.T.E.A. (Orchidaceae 3): 450. 1989.

Terrestrial herb. Montane grassland. 2640 m.

Distribution: KP—*Abeid 3081*

Eulophia elegans Schltr.; Cribb, F.T.E.A. (Orchidaceae 3): 446. 1989.

Terrestrial herb. Wet grassland at riverine forest edge. 1600 m.

Distribution: MP, RM—*Stolz 1980* (type: B, K, Z), *Kerfoot 1725*

Euolphia fridericii (Rchb. f.) A. V. Hall; Cribb, F.T.E.A. (Orchidaceae 3): 471. 1989.

Terrestrial herb. Montane grassland. 1940 m.

Distribution: SL—*Gereau 4672*

Eulophia guineensis Lindl.; Cribb, F.T.E.A. (Orchidaceae 3): 428. 1989.

Terrestrial herb. Abandoned cultivation. 1250 m.

Distribution: MP—*J. Lovett 3934*

Eulophia horsfallii (Bateman) Summerh.; Cribb, F.T.E.A. (Orchidaceae 3): 428. 1989.

Terrestrial herb. Riverine forest. 1850 m.

Distribution: RM, SL—*Gereau 4045*

Eulophia kyimbilae Schltr.; Cribb, F.T.E.A. (Orchidaceae 3): 446. 1989.

Terrestrial herb. Grassland, forest edges.

Distribution: MP, MR—*Cribb 11305*

Eulophia odontoglossa Rchb. f.; Cribb, F.T.E.A. (Orchidaceae 3): 458. 1989.

Terrestrial herb. Miombo woodland, *Cupressus* plantations, maize fields. 1540–1790 m.

Distribution: RM, SL—*Gereau 3878*

Eulophia streptopetala Lindl.; Cribb, F.T.E.A. (Orchidaceae 3): 469. 1989.

Epilithic herb. Riverine forest. 1500 m.

Distribution: SL—*Gereau 3946*

Habenaria clavata (Lindl.) Rchb. f.; Summerhayes, F.T.E.A. (Orchidaceae 1): 109. 1968.

Terrestrial herb. Woodland. 1250–1790 m.

Distribution: MP, SL—*Gereau 3890*

Habenaria cornuta Lindl.; Summerhayes, F.T.E.A. (Orchidaceae 1): 106. 1968.

Terrestrial herb. Montane grassland, cypress plantation. 1750–2400 m.

Distribution: KP, MP, RM—*Suleiman 51*

Habenaria goetzeana Kraenzl.; Summerhayes, F.T.E.A. (Orchidaceae 1): 114. 1968.

Terrestrial herb. Miombo woodland. 1790 m.

Distribution: SL—*Gereau 3888*

Habenaria isoantha Schltr.; Summerhayes, F.T.E.A. (Orchidaceae 1): 100. 1968.

Terrestrial herb. Grassland. 1600–1800 m.

Distribution: MR, RM—*Stolz 1148* (type: K, S), *Pettersson 472*

Habenaria kyimbilae Schltr.; Summerhayes, F.T.E.A. (Orchidaceae 1): 87. 1968.

Terrestrial herb. Montane grassland. 2230–2600 m.

Distribution: KP, SL—*Gereau 4436*

Habenaria leucoceras Schltr.; Summerhayes, F.T.E.A. (Orchidaceae 1): 90. 1968.

Terrestrial herb. Montane grassland. 2000 m.

Distribution: MP, NL—*Stolz 2490* (type: K), *Procter 1911*

Habenaria leucotricha Schltr.; Summerhayes, F.T.E.A. (Orchidaceae 1): 123. 1968.

Terrestrial herb. Woodland. 1200 m.

Distribution: MP—*J. Lovett 3875*

Habenaria lithophila Schltr.; Summerhayes, F.T.E.A. (Orchidaceae 1): 132. 1968.

Terrestrial herb. Rocky mountain slopes. 2800 m.

Distribution: RM—*Stolz 2505* (type: K)

Habenaria macrura Kraenzl.; Summerhayes, F.T.E.A. (Orchidaceae 1): 126. 1968.

Terrestrial herb. Montane grassland. 2400–2600 m.

Distribution: KP—*Suleiman 52*

Habenaria malacophylla Rchb. f.; Summerhayes, F.T.E.A. (Orchidaceae 1): 73. 1968.

Terrestrial herb. Montane forest. 2120–2600 m.

Distribution: MR, SL—*Gereau 4337*

Habenaria occlusa Summerh.; Summerhayes, F.T.E.A. (Orchidaceae 1): 67. 1968.

Terrestrial herb. Montane grassland. 2100–2590 m.

Distribution: KP, MP—*Richards 14178* (type: K), *Suleiman 20*

Habenaria odorata Schltr.; Summerhayes, F.T.E.A. (Orchidaceae 1): 121. 1968.

Terrestrial herb. Montane grassland. 2100 m.

Distribution: NL—*Stolz 2488* (type: K)

Habenaria petitiana (A. Rich.) T. Durand & Schinz; Summerhayes, F.T.E.A. (Orchidaceae 1): 56. 1968.

Terrestrial herb. Scree slopes, montane

grassland. 2060–2120 m.

Distribution: RM, SL—*Gereau 4447*

Habenaria praestans Rendle; Summerhayes, F.T.E.A. (Orchidaceae 1): 70. 1968.

Terrestrial herb. Riverine forest, miombo woodland, roadsides in cultivation. 1500–1900 m.

Distribution: MP, MR, NL, SL—*Gereau 4407*

Habenaria schimperiana Hochst. ex A. Rich.; Summerhayes, F.T.E.A. (Orchidaceae 1): 89. 1968.

Terrestrial herb. Woodland. 1200 m.

Distribution: MP—*J. Lovett 4243*

Habenaria uhehensis Schltr.; Summerhayes, F.T.E.A. (Orchidaceae 1): 77. 1968.

Terrestrial herb. Grassland. 2440 m.

Distribution: MP, RM—*Nicholson DSM2546*

Habenaria cf. *welwitschii* Rchb. f.; Summerhayes, F.T.E.A. (Orchidaceae 1): 83. 1968.

Terrestrial herb. Montane grassland. 2400 m.

Distribution: KP—*Suleiman 54*

Habenaria xanthochlora Schltr.; Summerhayes, F.T.E.A. (Orchidaceae): 56. 1968.

Terrestrial herb. Montane grassland. 2590–2600 m.

Distribution: KP—*Lovett 1859*

Holothrix nyasae Rolfe; Summerhayes, F.T.E.A. (Orchidaceae 1): 7. 1968.

Terrestrial herb. Montane grassland. 2400–2740 m.

Distribution: KP, MP, NL—*Suleiman 50*

Holothrix triloba (Rolfe) Kraenzl.; Summerhayes, F.T.E.A. (Orchidaceae 1): 11. 1968.

Terrestrial herb. Montane grassland. 2100–2250 m.

Distribution: RM—*Procter 1341*

Liparis bowkeri Harv.; Cribb & Bowden, F.T.E.A. (Orchidaceae 2): 299. 1984.

Terrestrial or epilithic herb. Miombo woodland, riverine forest, scree slopes. 1210–2120 m.

Distribution: SL—*Gereau 4406*

Liparis deistelii Schltr.; Cribb & Bowden, F.T.E.A. (Orchidaceae 2): 299. 1984.

Terrestrial or epiphytic herb. Forest.

Distribution: RM—*Richards 14352*

Liparis nervosa (Thunb.) Lindl.; Cribb & Bowden, F.T.E.A. (Orchidaceae 2): 298. 1984.

Terrestrial herb. Miombo woodland. 1530 m.

Distribution: SL—*Gereau 4370*

Liparis rungweensis Schltr.; Cribb & Bowden, F.T.E.A. (Orchidaceae 2): 297. 1984.

Terrestrial herb. Montane grassland. 1600–2960 m.

Distribution: MP, NL, RM—*Stolz 2438* (type: K), *Gereau 4089*

Liparis stolzii Schltr.; Cribb & Bowden, F.T.E.A. (Orchidaceae 2): 297. 1984.

Epiphyte. Forest. 1600 m.

Distribution: RM—*Stolz 1152* (type: K)

Malaxis schliebenii (Mansf.) Summerh.; Cribb, F.T.E.A. (Orchidaceae 2): 286. 1984.

Terrestrial herb. Miombo woodland. 1530–1790 m.

Distribution: SL—*Gereau 4366*

Margelliantha leedalii P. J. Cribb; Cribb, F.T.E.A. (Orchidaceae 3): 541. 1989.

Epiphyte. Montane forest. 2070 m.

Distribution: NL, RM—*Gereau 4139*

Mystacidium tanganyikense Summerh.; Cribb, F.T.E.A. (Orchidaceae 3): 594. 1989.

Epiphyte. Montane forest. 2070 m.

Distribution: MP, NL—*Gereau 4140*

Neobolusia stolzii Schltr.; Summerhayes, F.T.E.A. (Orchidaceae 1): 13. 1968.

Terrestrial herb. Grassland. 1850–2600 m.

Distribution: MR, RM—*Stolz 1076* (type: K), *Wingfield 173*

Platycoryne crocea (Schweinf. ex Rchb. f.) Rolfe subsp. *ochrantha* (Schltr.) Summerh.; Summerhayes, F.T.E.A. (Orchidaceae 1): 148. 1968.

Terrestrial herb. Miombo woodland. 1200 m.

Distribution: MP, NL—*J. Lovett 3991*

Polystachya albescens Ridl. subsp. *imbricata* (Rolfe) Summerh.; Cribb, F.T.E.A. (Orchidaceae 2): 355. 1984.

Terrestrial or epiphytic herb. Montane forest. 1750 m.

Distribution: RM—*Semsei 1594*

Polystachya cultriformis (Thouars) Lindl. ex

Spreng.; Cribb, F.T.E.A. (Orchidaceae 2): 341. 1984.

Epiphyte. Montane forest, riverine forest. 2000 m.

Distribution: MP, RM—*J. Lovett 4987*

Polystachya fusiformis (Thouars) Lindl.; Cribb, F.T.E.A. (Orchidaceae 2): 371. 1984.

Epiphyte. Montane forest. 2070 m.

Distribution: NL—*Gereau 4138*

Polystachya malilaensis Schltr.; Cribb, F.T.E.A. (Orchidaceae 2): 398. 1984.

Epiphytic or epilithic herb. Montane forest, riverine forest. 1850–1980 m.

Distribution: MP, SL—*Gereau 4037*

Polystachya modesta Rchb. f.; Cribb, F.T.E.A. (Orchidaceae 2): 363. 1984.

Epiphyte. Riverine forest. 1210 m.

Distribution: SL—*Gereau 4266*

Pterygodium ukingense Schltr.; Hunt, F.T.E.A. (Orchidaceae 1): 216. 1968.

Terrestrial herb. Montane grassland. 2000 m.

Distribution: NL—*Stolz 2609* (type: K)

Satyrium acutirostrum Summerh.; Summerhayes, F.T.E.A. (Orchidaceae 1): 196. 1968.

Terrestrial herb. Montane grassland. 1950–2610 m.

Distribution: KP, MP, RM—*Suleiman 21*

Satyrium anomalum Schltr.; Summerhayes, F.T.E.A. (Orchidaceae 1): 205. 1968.

Terrestrial herb. Miombo woodland. 1790 m.

Distribution: SL—*Gereau 3887*

Satyrium buchananii Schltr.; Summerhayes, F.T.E.A. (Orchidaceae 1): 199. 1968.

Terrestrial herb. Montane grassland, riverine forest, miombo woodland. 1790–2700 m.

Distribution: KP, NL, RM, SL—*Suleiman 65*

Satyrium chlorocorys Rchb. f. ex Rolfe; Summerhayes, F.T.E.A. (Orchidaceae 1): 206. 1968.

Terrestrial herb. Montane grassland, miombo woodland. 1610–2400 m.

Distribution: KP, MP, RM, SL—*Gereau 3866*

Satyrium comptum Summerh.; Summerhayes, F.T.E.A. (Orchidaceae 1): 208. 1968.

Terrestrial herb. Cracks in rock face. 2650–2800 m.

Distribution: MP, RM—*Stolz 2504* (type: K), *Wingfield 936*

Satyrium crassicaule Rendle; Summerhayes, F.T.E.A. (Orchidaceae 1): 193. 1968.

Terrestrial herb. Riverbanks and marshes in montane grassland. 1860–2600 m.

Distribution: KP, MP, SL—*Gereau 3863*

Satyrium longicauda Lindl.; Summerhayes, F.T.E.A. (Orchidaceae 1): 198. 1968.

Terrestrial herb. Montane grassland. 2610 m.

Distribution: KP, MP—*Kerfoot 1607*

Satyrium microcorys Schltr.; Summerhayes, F.T.E.A. (Orchidaceae 1): 207. 1968.

Terrestrial herb. Rocky montane grassland. 2000–2570 m.

Distribution: KP, MP, RM—*Stolz 2436* (type: K), *Gereau 3165*

Satyrium miserum Kraenzl. vel sp. aff.; Summerhayes, F.T.E.A. (Orchidaceae 1): 207. 1968.

Terrestrial herb. Rocky montane grassland. 2730 m.

Distribution: KP—*Abeid 3079*

Satyrium monadenum Schltr.; Summerhayes, F.T.E.A. (Orchidaceae 1): 195. 1968.

Terrestrial herb. Montane grassland. 2400–2840 m.

Distribution: KP, RM—*Stolz 1031* (type: K), *Abeid 2785*

Satyrium neglectum Schltr.; Summerhayes, F.T.E.A. (Orchidaceae 1): 197. 1968.

Terrestrial herb. Montane grassland, disturbed shrubland. 1850–2740 m.

Distribution: KP, MP, NL, RM, SL—*Gereau 4231*

Satyrium princeae Kraenzl.; Summerhayes, F.T.E.A. (Orchidaceae 1): 189. 1968.

Terrestrial herb. Montane grassland. 2400–2550 m.

Distribution: KP, RM, SL—*Suleiman 30*

Satyrium rhynchantoides Schltr.; Summerhayes, F.T.E.A. (Orchidaceae 1): 195. 1968.

Terrestrial herb. Montane grassland. 2400 m.

Distribution: RM—*Stolz 1074, 2435* (syntypes: K)

Satyrium sceptrum Schltr.; Summerhayes, F.T.E.A. (Orchidaceae 1): 197. 1968.

Terrestrial herb. Montane grassland. 2000–2570 m.

Distribution: KP, RM—*Stolz 2434* (type: EA, K), *Gereau 3091*

Satyrium sphaeranthum Schltr.; Summerhayes, F.T.E.A. (Orchidaceae 1): 208. 1968.

Terrestrial herb. Montane grassland, montane forest. 2080–2600 m.

Distribution: KP, MP, RM, SL—*Suleiman 111*

Satyrium trinerve Lindl.; Hall, 1999: 155.

Terrestrial herb. Montane grassland, miombo woodland. 1790–1860 m.

Distribution: SL—*Gereau 3882*

Satyrium volkensii Schltr.; Summerhayes, F.T.E.A. (Orchidaceae 1): 202. 1968.

Terrestrial herb. Woodland, montane grassland. 1300–1860 m.

Distribution: MP, SL—*Gereau 3879*

Schizochilus sulphureus Schltr.; Summerhayes, F.T.E.A. (Orchidaceae 1): 27. 1968.

Terrestrial herb. Montane grassland. 2400–2750 m.

Distribution: KP, MP, RM—*Stolz 1075* (type: K), *Kerfoot 1631*

Stenoglottis fimbriata Lindl.; Summerhayes, F.T.E.A. (Orchidaceae 1): 29. 1968.

Epiphyte. Montane forest and openings. 1800 m.

Distribution: NL, RM—*Bidgood 98*

Ypsilopus erectus (P. J. Cribb) P. J. Cribb & J. L. Stewart; Cribb, F.T.E.A. (Orchidaceae 3): 628. 1989.

Epiphyte. Woodland.

Distribution: RM—*Moreau 871*

OROBANCHACEAE

1 genus and 1 species

Orobanche minor Sm.; Graham, F.T.E.A. (Orobanchaceae): 5. 1957

Parasitic terrestrial herb. Woodland, riverine forest, montane forest and shrubland, abandoned cultivation. 1250–2500 m.

Distribution: MP, SL—*Suleiman 182*

OXALIDACEAE

2 genera and 4 species

Biophytum abyssinicum Steud. ex A. Rich.; Kabuye, F.T.E.A. (Oxalidaceae): 12. 1971.

Terrestrial herb. Woodland. 1250 m.

Distribution: MP—*J. Lovett 4073*

Biophytum umbraculum Welw.; Lourteig, 1981: 451.

Terrestrial herb. Fallow fields. 520–1620 m.

Distribution: RM, SL—*Mboya 241*

Oxalis corniculata L.; Kabuye, F.T.E.A. (Oxalidaceae): 3. 1971.

Terrestrial herb. Montane forest.

Distribution: MP, RM—*Bullock 3987*

Oxalis obliquifolia Steud. ex A. Rich.; Kabuye, F.T.E.A. (Oxalidaceae): 6. 1971.

Terrestrial herb. Rocky mountain summit, montane grassland. 2590–2600 m.

Distribution: KP, MP—*Gereau 3500*

PASSIFLORACEAE

2 genera and 3 species

Adenia lindiensis Harms; Wilde, F.T.E.A. (Passifloraceae): 39. 1975.

Herbaceous climber. Riverine forest. 580 m.

Distribution: SL—*Gereau 3814*

Adenia stenodactyla Harms; Wilde, F.T.E.A. (Passifloraceae): 37. 1975.

Herbaceous climber. Woodland, montane thickets. 1300–1920 m.

Distribution: MP, SL—*Gereau 3609*

Passiflora edulis Sims; Wilde, F.T.E.A. (Passifloraceae): 15. 1975.

Herbaceous climber. Montane forest edges. 1720–1850 m.

Distribution: RM—*Bracebridge 91*

PEDALIACEAE

1 genus and 2 species

Sesamum angolense Welw.; Bruce, F.T.E.A. (Pedaliaceae): 20. 1953.

Herb or subshrub. Woodland. 1200–1460 m.

Distribution: MP—*Kayombo 2428*

Sesamum angustifolium (Oliv.) Engl.; Bruce,

F.T.E.A. (Pedaliaceae): 19. 1953.

Terrestrial herb. Woodland, forest edges. 1200–1520 m.

Distribution: MP, RM—*J. Lovett 4242*

PIPERACEAE

2 genera and 5 species

Peperomia fernandopoiana C. DC. var. *butaguensis* (De Wild.) Düll.; Verdcourt, F.T.E.A. (Piperaceae): 21. 1996.

Epiphyte. Montane forest. 1750 m.

Distribution: RM—*Kindeketa 2972*

Note: New to Tanzania, from Kenya, Uganda, Democratic Republic of Congo, and Rwanda.

Peperomia retusa (L. f.) A. Dietr.; Verdcourt, F.T.E.A. (Piperaceae): 13. 1996.

Epiphyte. Montane forest. 1780 m.

Distribution: RM—*Bracebridge 99A*

Peperomia tetraphylla (G. Forst.) Hook. & Arn.; Verdcourt, F.T.E.A. (Piperaceae): 12. 1996.

Epiphyte. Montane forest. 1780 m.

Distribution: RM—*Bracebridge 99*

Piper capense L. f.; Verdcourt, F.T.E.A. (Piperaceae): 5. 1996.

Shrub. Montane forest, tree clumps in grassland. 1720–2460 m.

Distribution: MP, MR, NL, RM, SL—*Gereau 4713*

Piper umbellatum L.; Verdcourt, F.T.E.A. (Piperaceae): 8. 1996.

Shrub. Riverine forest. 1160–1210 m.

Distribution: SL—*Gereau 4260*

PITTOSPORACEAE

1 genus and 2 species

Pittosporum abyssinicum Delile; Lovett et al., 2006: 201.

Tree. Montane forest edges, tree clumps in montane grassland. 2410–2700 m.

Distribution: KP, NL, RM—*Kibure 1412*

Pittosporum viridiflorum Sims; Lovett et al., 2006: 201.

Tree or shrub. Montane forest, riverine forest, miombo woodland, lakeshore shrubland. 474–2470 m.

Distribution: KP, MR, SL—*Gereau 4271*

PLANTAGINACEAE

1 genus and 2 species

Plantago lanceolata L.; Verdcourt, F.T.E.A. (Plantaginaceae): 6. 1971.

Terrestrial herb. Montane grassland. 2520 m.

Distribution: KP—*WCS SHCP 407*

Plantago palmata Hook. f.; Verdcourt, F.T.E.A. (Plantaginaceae): 2. 1971.

Terrestrial herb. Montane forest. 2015 m.

Distribution: MR, RM—*Gereau 7010*

PLUMBAGINACEAE

1 genus and 1 species

Plumbago zeylanica L.; Wilmot-Dear, F.T.E.A. (Plumbaginaceae): 6. 1976.

Shrub. Montane forest edges. 1760 m.

Distribution: RM—*WCS SHCP 171*

POACEAE

50 genera and 124 species

Agrostis continuata Stapf; Clayton, F.T.E.A. (Gramineae 1): 111. 1970.

Terrestrial herb. Stream banks in montane grassland. 2560 m.

Distribution: KP—*Gereau 3113*

Agrostis kilimandscharica Mez; Clayton, F.T.E.A. (Gramineae 1): 110. 1970.

Terrestrial herb. Montane grassland, seasonal swamps, rock face near waterfall. 2570–2920 m.

Distribution: KP, MP, RM—*Gereau 3082*

Agrostis lachnantha Nees; Clayton, F.T.E.A. (Gramineae 1): 106. 1970.

Terrestrial herb. Open woodland. 1650–1680 m.

Distribution: SL—*Gereau 5113*

Andropogon amethystinus Steud.; Clayton & Renvoize, F.T.E.A. (Gramineae 3): 772. 1982.

Terrestrial herb. Road beds in cultivation, montane grassland. 2200–2840 m.

Distribution: KP, MR—*Abeid 2853*

Andropogon chinensis (Nees) Merr.; Clayton & Renvoize, F.T.E.A. (Gramineae 3): 779. 1982.

Terrestrial herb. Riverine forest. 1500 m.

Distribution: SL—*Gereau 4806*

Andropogon chrysostachyus Steud.; Clayton & Renvoize, F.T.E.A. (Gramineae 3): 773. 1982.

Terrestrial herb. Montane grassland. 2525–2770 m.

Distribution: KP, RM—*Abeid 2831*

Andropogon distachyos L.; Clayton & Renvoize, F.T.E.A. (Gramineae 3): 770. 1982.

Terrestrial herb. Open and disturbed places in montane forest, montane grassland. 2200–2300 m.

Distribution: MR—*Gereau 4616*

Andropogon mannii Hook. f.; Clayton & Renvoize, F.T.E.A. (Gramineae 3): 774. 1982.

Terrestrial herb. Montane grassland, stream banks. 2440–2740 m.

Distribution: KP, MR, RM—*Gereau 3170*

Andropogon schirensis Hochst. ex A. Rich.; Clayton & Renvoize, F.T.E.A. (Gramineae 3): 779. 1982.

Terrestrial herb. Montane grassland, woodland. 1260–2320 m.

Distribution: MP, SL—*Suleiman 162*

Andropterum stolzii (Pilg.) C. E. Hubb.; Clayton & Renvoize, F.T.E.A. (Gramineae 3): 752. 1982.

Terrestrial herb. Open woodland, montane grassland. 1650–2300 m.

Distribution: MR, SL—*Gereau 4762*

Arthraxon micans (Nees) Hochst.; Clayton & Renvoize, F.T.E.A. (Gramineae 3): 742. 1982.

Terrestrial herb. Montane grassland, road banks. 1940–2200 m.

Distribution: MR, SL—*Gereau 4776*

Arundinella nepalensis Trin.; Hubbard, F.T.E.A. (Gramineae 2): 409. 1974.

Terrestrial herb. Eucalyptus woodland. 1250 m.

Distribution: MP—*J. Lovett 4425*

Bewsia biflora (Hack.) Gooss.; Clayton, F.T.E.A. (Gramineae 2): 286. 1974.

Terrestrial herb. Miombo woodland. 1260–1530 m.

Distribution: SL—*Gereau 4361*

Bothriochloa insculpta (Hochst. ex A. Rich.) A. Camus; Clayton & Renvoize, F.T.E.A. (Gramineae 3): 720. 1982.

Terrestrial herb. Roadsides in miombo woodland. 1500 m.

Distribution: SL—*Gereau 3955*

Brachiaria decumbens Stapf; Clayton & Renvoize, F.T.E.A. (Gramineae 3): 586. 1982.

Terrestrial herb. Abandoned cultivation. 1250 m.

Distribution: MP—*J. Lovett 4543*

Brachiaria deflexa (Schumach.) C. E. Hubb. ex Robyns; Clayton & Renvoize, F.T.E.A. (Gramineae 3): 598. 1982.

Terrestrial herb. Roadsides. 474–490 m.

Distribution: SL—*Gereau 3789*

Brachiaria jubata (Fig. & De Not.) Stapf; Clayton & Renvoize, F.T.E.A. (Gramineae 3): 580. 1982.

Terrestrial herb. Woodland. 1250–1950 m.

Distribution: MP—*J. Lovett 3519*

Brachiaria reptans (L.) C. A. Gardner & C. E. Hubb.; Clayton & Renvoize, F.T.E.A. (Gramineae 3): 591. 1982.

Terrestrial herb. Miombo woodland. 1200 m.

Distribution: MP—*J. Lovett 4320*

Brachiaria scalaris Pilg.; Clayton & Renvoize, F.T.E.A. (Gramineae 3): 594. 1982.

Terrestrial herb. Woodland. 2000 m.

Distribution: MP—*P. Lovett 268*

Brachiaria serrata (Thunb.) Stapf; Clayton & Renvoize, F.T.E.A. (Gramineae 3): 588. 1982.

Terrestrial herb. Miombo woodland. 1200–1790 m.

Distribution: MP, SL—*Gereau 3898*

Brachypodium flexum Nees; Clayton, F.T.E.A. (Gramineae 1): 71. 1970.

Scrambling terrestrial herb. Montane forest, stream banks in montane grassland. 2040–2570 m.

Distribution: KP, SL—*Gereau 4728*

Briza minor L.; Hubbard, F.T.E.A. (Gramineae 1): 53. 1970.

Terrestrial herb. Cultivated and waste ground. 2000 m.

Distribution: RM—*Semsei 1570*

Bromus leptoclados Nees; Clayton, F.T.E.A. (Gramineae 1): 68. 1970.

Terrestrial herb. Montane grassland. 2360–2570 m.

Distribution: KP, MR—*Gereau 3777*

Cymbopogon nardus (L.) Rendle; Clayton & Renvoize, F.T.E.A. (Gramineae 3): 764. 1982.

Terrestrial herb. Miombo woodland. 1260–1500 m.

Distribution: SL—*Gereau 4413*

Cynodon nlemfuensis Vanderyst; Clayton, F.T.E.A. (Gramineae 2): 319. 1974.

Terrestrial herb. Seasonally inundated grassland. 1270 m.

Distribution: MP—*Gereau 3490*

Digitaria abyssinica (Hochst. ex A. Rich.) Stapf; Clayton & Renvoize, F.T.E.A. (Gramineae 3): 641. 1982.

Terrestrial herb. Montane grassland, valley grassland. 1270–2360 m.

Distribution: MR, SL—*Suleiman 157*

Digitaria diagonalis (Nees) Stapf var. *hirsuta* (De Wild. & T. Durand) Stapf ex Troupin; Clayton & Renvoize, F.T.E.A. (Gramineae 3): 626. 1982.

Terrestrial herb. Miombo woodland. 1790 m.

Distribution: SL—*Gereau 3897*

Digitaria diagonalis (Nees) Stapf var. *uniglumis* (Hochst. ex A. Rich.) Pilg.; Clayton & Renvoize, F.T.E.A. (Gramineae 3): 626. 1982.

Terrestrial herb. Miombo woodland. 1200–1530 m.

Distribution: MP, SL—*Gereau 4384*

Digitaria gazensis Rendle; Clayton & Renvoize, F.T.E.A. (Gramineae 3): 643. 1982.

Terrestrial herb. Roadsides in miombo woodland. 1500 m.

Distribution: SL—*Gereau 3956*

Digitaria myurus Stapf; Clayton & Renvoize, F.T.E.A. (Gramineae 3): 627. 1982.

Terrestrial herb. Montane grassland. 2100–2700 m.

Distribution: KP, MP, SL—*Stolz 2619* (type: K, US), *Kerfoot 1609*

Digitaria pearsonii Stapf; Clayton & Renvoize, F.T.E.A. (Gramineae 3): 643. 1982.

Terrestrial herb. Montane grassland, clearings in montane forest, scree slopes, woodland. 1900–2230 m.

Distribution: NL, SL—*Gereau 4442*

Digitaria perrottetii (Kunth) Stapf; Clayton & Renvoize, F.T.E.A. (Gramineae 3): 649. 1982.

Terrestrial herb. Cultivated areas, woodland. 1250–1500 m.

Distribution: MP—*Kayombo 2343*

Digitaria rukwae Clayton; Clayton & Renvoize, F.T.E.A. (Gramineae 3): 646. 1982.

Terrestrial herb. Trails and edges of cultivation in montane grassland. 2470–2590 m.

Distribution: KP—*Gereau 3127*

Digitaria scalarum (Schweinf.) Chiov.; Goetghebeur & Van der Veken, Fl. Zamb. 10(3): 159. 1989.

Terrestrial herb. Seasonal pond margins in montane grassland, agricultural fields. 1840–2300 m.

Distribution: MR, SL—*Gereau 3911*

Digitaria ternata (Hochst. ex A. Rich.) Stapf; Clayton & Renvoize, F.T.E.A. (Gramineae 3): 630. 1982.

Terrestrial herb. Montane grassland, road beds, woodland, abandoned cultivation. 1250–2200 m.

Distribution: MP, MR, SL—*Gereau 4777*

Digitaria thouarsiana (Flüggé) A. Camus; Clayton & Renvoize, F.T.E.A. (Gramineae 3): 631. 1982.

Terrestrial herb. Clearings in montane forest. 1900 m.

Distribution: NL—*Gereau 4192*

Eleusine coracana (L.) Gaertn. subsp. *africana* (Kenn.-O'Byrne) Hilu & de Wet; Hilu, 1994: 411.

Terrestrial herb. Roadsides. 474–490 m.

Distribution: SL—*Gereau 3791*

Elionurus muticus (Spreng.) Kuntze; Clayton & Renvoize, F.T.E.A. (Gramineae 3): 837. 1982.

Terrestrial herb. Montane grassland. 2250 m.

Distribution: MP—*J. Lovett 4859*

Eragrostis capensis (Thunb.) Trin.; Clayton, F.T.E.A. (Gramineae 2): 221. 1974.

Terrestrial herb. Montane grassland. 1940 m.

Distribution: SL—*Gereau 4768*

Eragrostis congesta Oliv.; Clayton, F.T.E.A. (Gramineae 2): 224. 1974.

Terrestrial herb. Montane grassland. 1940 m.

Distribution: SL—*Gereau 4766*

Eragrostis exasperata Peter; Clayton, F.T.E.A. (Gramineae 2): 220. 1974.

Terrestrial herb. Roadsides. 474–490 m.

Distribution: SL—*Gereau 3798*

Eragrostis hispida K. Schum.; Clayton, F.T.E.A. (Gramineae 2): 201. 1974.

Terrestrial herb. Clearings in riverine forest, woodland. 1250–1850 m.

Distribution: MP, SL—*Gereau 4052*

Eragrostis macilenta (A. Rich.) Steud.; Clayton, F.T.E.A. (Gramineae 2): 235. 1974.

Terrestrial herb. Riverine forest, edges of cultivation. 1500–1700 m.

Distribution: MP, SL—*Gereau 3959*

Eragrostis mollior Pilg. ex R. E. Fr.; Clayton, F.T.E.A. (Gramineae 2): 202. 1974.

Terrestrial herb. Miombo woodland. 1530 m.

Distribution: SL—*Gereau 4383*

Eragrostis racemosa (Thunb.) Steud.; Clayton, F.T.E.A. (Gramineae 2): 230. 1974.

Terrestrial herb. Montane grassland, roadsides in villages. 1360–2700 m.

Distribution: KP, SL—*Suleiman 164*

Eragrostis schweinfurthii Chiov. var. *schweinfurthii*; Phillips, 1995: 117.

Terrestrial herb. Montane grassland, edges of cultivation, road beds. 1940–2590 m.

Distribution: KP, MR, SL—*Gereau 4769*

Eragrostis setulifera Pilg.; Clayton, F.T.E.A. (Gramineae 2): 205. 1974.

Terrestrial herb. Disturbed cultivated areas. 1250 m.

Distribution: MP—*Kayombo 2465*

Eragrostis tenuifolia (A. Rich.) Hochst. ex Steud.; Clayton, F.T.E.A. (Gramineae 2): 238. 1974.

Terrestrial herb. Montane grassland, abandoned cultivation. 1250–2150 m.

Distribution: MP, SL—*Suleiman 159*

Eulalia villosa Nees; Clayton & Renvoize, F.T.E.A. (Gramineae 3): 713. 1982.

Terrestrial herb. Montane grassland. 2140–2150 m.

Distribution: SL—*Suleiman 152*

Exotheca abyssinica (Hochst. ex A. Rich.) Andersson; Clayton & Renvoize, F.T.E.A. (Gramineae 3): 821. 1982.

Terrestrial herb. Montane grassland, openings in montane forest, roadsides in cultivation. 2200–2860 m.

Distribution: KP, MR, RM—*Abeid 2859*

Festuca abyssinica Hochst. ex A. Rich. var. *abyssinica*; Clayton, F.T.E.A. (Gramineae 1): 60. 1970.

Terrestrial herb. Montane grassland. 2560–2840 m.

Distribution: KP—*Abeid 2852*

Festuca africana (Hack.) Clayton; Clayton, 1985: 727.

Terrestrial herb. Montane forest. 1700–2130 m.

Distribution: MR, RM—*Abeid 2931*

Festuca caprina Nees; Clayton, F.T.E.A. (Gramineae 1): 62. 1970.

Terrestrial herb. Montane grassland. 2560–2660 m.

Distribution: KP, MP, NL—*Abeid 2803*

Festuca engleri Pilg.; Phillips, 1995: 27.

Terrestrial herb. Cypress plantation. 2100 m.

Distribution: MP, RM—*Geilinger 2337*

Helictotrichon elongatum (Hochst. ex A. Rich.) C. E. Hubb.; Clayton, F.T.E.A. (Gramineae 1): 89. 1970.

Terrestrial herb. Montane grassland. 1940 m.

Distribution: SL—*Gereau 4764*

Helictotrichon milanjianum (Rendle) C. E. Hubb.; Clayton, F.T.E.A. (Gramineae 1): 88. 1970.

Terrestrial herb. Clearings in montane forest. 2360 m.

Distribution: NL—*Gereau 4126*

Helictotrichon umbrosum (Hochst. ex Steud.) C. E. Hubb.; Clayton, F.T.E.A. (Gramineae 1): 88. 1970.

Terrestrial herb. Montane grassland. 2570 m.

Distribution: KP—*Gereau 3108*

Hyparrhenia anamesa Clayton; Clayton & Renvoize, F.T.E.A. (Gramineae 3): 800. 1982.

Terrestrial herb. Montane grassland. 1940 m.

Distribution: SL—*Gereau 4753*

Hyparrhenia cymbaria (L.) Stapf; Clayton & Renvoize, F.T.E.A. (Gramineae 3): 804. 1982.

Terrestrial herb. Montane grassland, openings in montane forest, cultivated areas. 1250–2200 m.

Distribution: MP, MR, SL—*Kayombo 2341*

Hyparrhenia dregeana (Nees) Stapf ex Stent; Clayton & Renvoize, F.T.E.A. (Gramineae 3): 809. 1982.

Terrestrial herb. Montane grassland near forest edge. 2560 m.

Distribution: KP—*Abeid 2872*

Hyparrhenia filipendula (Hochst.) Stapf; Clayton & Renvoize, F.T.E.A. (Gramineae 3): 803. 1982.

Terrestrial herb. Riverine forest edges. 1250–1500 m.

Distribution: MP, SL—*Gereau 4412*

Hyparrhenia nyassae (Rendle) Stapf; Clayton & Renvoize, F.T.E.A. (Gramineae 3): 793. 1982.

Terrestrial herb. Riverine forest. 1500 m.

Distribution: SL—*Gereau 4807*

Hyparrhenia pilgeriana C. E. Hubb.; Clayton & Renvoize, F.T.E.A. (Gramineae 3): 807. 1982.

Terrestrial herb. Montane shrubland. 1560–2300 m.

Distribution: MR, RM—*Mwasumbi 16301*

Hyparrhenia quarrei Robyns; Clayton & Renvoize, F.T.E.A. (Gramineae 3): 799. 1982.

Terrestrial herb. Montane grassland near forest edge. 2540 m.

Distribution: KP—*Abeid 2865*

Hyparrhenia rufa (Nees) Stapf; Clayton & Renvoize, F.T.E.A. (Gramineae 3): 794. 1982.

Terrestrial herb. Riverine forest, miombo woodland, abandoned cultivation. 1250–1800 m.

Distribution: MP, SL—*Gereau 4808*

Isachne mauritiana Kunth; Clayton, F.T.E.A. (Gramineae 2): 434. 1974.

Terrestrial herb. Clearings in montane forest. 1790 m.

Distribution: NL—*Gereau 4152*

Koeleria capensis Nees; Clayton, F.T.E.A. (Gramineae 1): 79. 1970.

Terrestrial herb. Montane grassland. 2040–2800 m.

Distribution: KP, MP, MR, RM—*Stolz 2517A*

Loudetia simplex (Nees) C. E. Hubb.; Clayton, F.T.E.A. (Gramineae 2): 418. 1974.

Terrestrial herb. Montane grassland. 2140–2150 m.

Distribution: SL—*Suleiman 163*

Melinis minutiflora P. Beauv.; Zizka, 1988: 92.

Terrestrial herb. Montane grassland. 1940–2530 m.

Distribution: KP, SL—*Abeid 2866*

Melinis repens (Willd.) Zizka; Zizka, 1988: 55.

Terrestrial herb. Woodland. 1250 m.

Distribution: MP—*J. Lovett 4542*

Melinis scabrida (K. Schum.) Hack.; Zizka, 1988: 80.

Terrestrial herb. Montane grassland. 2140–2150 m.

Distribution: SL—*Suleiman 165A*

Microchloa caffra Nees; Renvoize, F.T.E.A. (Gramineae 2): 316. 1974.

Terrestrial herb. Montane grassland. 2320 m.

Distribution: SL—*Gereau 3989*

Microchloa kunthii Desv.; Renvoize, F.T.E.A. (Gramineae 2): 314. 1974.

Terrestrial herb. Montane grassland, valley grassland. 1270–2570 m.

Distribution: KP, MP—*Gereau 3479*

Oplismenus burmannii (Retz.) P. Beauv.; Clayton & Renvoize, F.T.E.A. (Gramineae 3): 542. 1982.

Terrestrial herb. Woodland, riverine forest. 1250 m.

Distribution: MP—*J. Lovett 4480*

Oplismenus compositus (L.) P. Beauv.; Clayton & Renvoize, F.T.E.A. (Gramineae 3): 542. 1982.

Terrestrial herb. Riverine forest. 1275–1600 m.

Distribution: MP—*Kayombo 2363*

Oplismenus hirtellus (L.) P. Beauv.; Clayton & Renvoize, F.T.E.A. (Gramineae 3): 542. 1982.

Terrestrial herb. Montane forest. 1710–2110 m.

Distribution: MR, RM, SL—*Gereau 4736*

Oplismenus undulatifolius (Ard.) P. Beauv.; Clayton & Renvoize, F.T.E.A. (Gramineae 3): 544. 1982.

Terrestrial herb. Montane forest.

Distribution: RM—*St. Clair-Thompson 1009*

Oxytenanthera abyssinica (A. Rich.) Munro; Clayton, F.T.E.A. (Gramineae 1): 11. 1970.

Bamboo. Villages. 2000 m.

Distribution: NL—*Wiland 152*

Panicum adenophorum K. Schum.; Clayton & Renvoize, F.T.E.A. (Gramineae 3): 469. 1982.

Scrambling terrestrial herb. Montane thickets. 1840 m.

Distribution: SL—*Gereau 3544*

Panicum calvum Stapf; Clayton & Renvoize, F.T.E.A. (Gramineae 3): 493. 1982.

Scrambling terrestrial herb. Montane forest. 2040 m.

Distribution: SL—*Gereau 4722*

Panicum claytonii Renvoize; Clayton & Renvoize, F.T.E.A. (Gramineae 3): 467. 1982.

Scrambling terrestrial herb. Montane forest, forest edges, scree slopes. 1980–2230 m.

Distribution: NL, SL—*Gereau 4134*

Panicum delicatulum Fig. & De Not.; Clayton & Renvoize, F.T.E.A. (Gramineae 3): 491. 1982.

Terrestrial herb. Riverine forest, road banks. 1500–2200 m.

Distribution: MR, SL—*Gereau 4529*

Panicum ecklonii Nees; Clayton & Renvoize, F.T.E.A. (Gramineae 3): 466. 1982.

Terrestrial herb. Rocky mountain slopes, edges of cultivation. 2470–2800 m.

Distribution: KP—*Gereau 3123*

Panicum gracilicaule Rendle; Clayton & Renvoize, F.T.E.A. (Gramineae 3): 474. 1982.

Terrestrial herb. Woodland. 1350 m.

Distribution: MP—*J. Lovett 4554*

Panicum haplocaulos Pilg.; Clayton & Renvoize, F.T.E.A. (Gramineae 3): 487. 1982.

Terrestrial herb. Seasonally waterlogged grassland. 1250 m.

Distribution: MP—*J. Lovett 4224*

Panicum inaequilatum Stapf & C. E. Hubb.; Renvoize, Fl. Zamb. 10(3): 38. 1989.

Terrestrial herb. Montane grassland, edges of cultivation. 2230–2470 m.

Distribution: KP, SL—*Gereau 3125*

Note: New to East Africa, from Mozambique, Malawi, and Zimbabwe.

Panicum issongense Pilg.; Clayton & Renvoize, F.T.E.A. (Gramineae 3): 495. 1982.

Terrestrial herb. Montane forest. 2180 m.

Distribution: MR, RM—*Mwasumbi 16195*

Panicum lukwangulense Pilg.; Clayton & Renvoize, F.T.E.A. (Gramineae 3): 467. 1982.

Terrestrial herb. Montane grassland and thicket edges. 2290–2740 m.

Distribution: KP, MR, RM—*Gereau 3124*

Panicum maximum Jacq.; Clayton & Renvoize, F.T.E.A. (Gramineae 3): 471. 1982.

Terrestrial herb. Riverine forest, abandoned cultivation. 1200–1250 m.

Distribution: MP, SL—*Gereau 4295*

Panicum mitopus K. Schum.; Clayton & Renvoize, F.T.E.A. (Gramineae 3): 498. 1982.

Terrestrial herb. Riverbanks in montane forest. 1880 m.

Distribution: NL—*Gereau 4225*

Panicum trichocladum Hack. ex K. Schum.; Clayton & Renvoize, F.T.E.A. (Gramineae 3): 473. 1982.

Terrestrial herb. Riverine forest. 1600 m.

Distribution: MP—*P. Lovett 225*

Paspalum scrobiculatum L.; Clayton & Renvoize, F.T.E.A. (Gramineae 3): 610. 1982.

Terrestrial herb. Montane grassland, roadsides in miombo woodland, abandoned cultivation. 1250–2150 m.

Distribution: MP, SL—*Suleiman 160*

Pennisetum clandestinum Hochst. ex Chiov.; Clayton & Renvoize, F.T.E.A. (Gramineae 3): 675. 1982.

Terrestrial herb. Roadsides in montane grassland. 2600 m.

Distribution: KP—*Gereau 3205*

Pennisetum macrourum Trin.; Clayton & Renvoize, F.T.E.A. (Gramineae 3): 689. 1982.

Terrestrial herb. Riverine forest, montane forest edges. 1500–2360 m.

Distribution: MR, SL—*Gereau 4805*

Pennisetum purpureum Schumach.; Clayton & Renvoize, F.T.E.A. (Gramineae 3): 677. 1982.

Terrestrial herb. Grassland, cultivated and abandoned fields, riverine forest. 1150–1400 m.

Distribution: MP—*Kayombo 2342*

Pennisetum sphacelatum (Nees) T. Durand & Schinz; Clayton & Renvoize, F.T.E.A. (Gramineae 3): 689. 1982.

Terrestrial herb. Montane grassland near forest edge. 2530 m.

Distribution: KP—*Abeid 2869*

Pennisetum thunbergii Kunth; Clayton & Renvoize, F.T.E.A. (Gramineae 3): 687. 1982.

Terrestrial herb. Montane grassland. 2080–2140 m.

Distribution: SL—*Suleiman 109*

Pennisetum unisetum (Nees) Benth.; Clayton & Renvoize, F.T.E.A. (Gramineae 3): 681. 1982.

Terrestrial herb. Woodland, riverine forest. 1250 m.

Distribution: MP—*J. Lovett 4541*

Pentaschistis natalensis Stapf; Clayton, F.T.E.A. (Gramineae 1): 125. 1970.

Terrestrial herb. Montane grassland and heathland. 2100–2880 m.

Distribution: KP, MP, MR, RM—*Abeid 3001*

Phragmites mauritianus Kunth; Clayton, F.T.E.A. (Gramineae 1): 120. 1970.

Terrestrial herb. Wet meadows in miombo woodland, riverine forest. 840–1980 m.

Distribution: MP, SL—*P. Lovett 276*

Poa schimperiana Hochst. ex A. Rich.; Clayton, F.T.E.A. (Gramineae 1): 48. 1970.

Terrestrial herb. Montane grassland. 2570 m.

Distribution: KP, MR—*Gereau 3085*

Poecilostachys oplismenoides (Hack.) Clayton; Clayton, Fl. Zamb. 10(3): 3. 1989.

Terrestrial herb. Riverine forest, montane forest. 1560–1780 m.

Distribution: RM—*Mwasumbi 16299*

Polypogon schimperianus (Hochst. ex Steud.) Cope; Phillips, 1995: 45.

Terrestrial herb. Riverine forest, stream banks in montane grassland. 1500–2590 m.

Distribution: KP, MP—*Gereau 2845*

Rottboellia cochinchinensis (Lour.) Clayton; Clayton & Renvoize, F.T.E.A. (Gramineae 3): 853. 1982.

Terrestrial herb. Woodland. 1300–1400 m.

Distribution: MP—*Kayombo 1009*

Sacciolepis transbarbata Stapf; Clayton & Renvoize, F.T.E.A. (Gramineae 3): 456. 1982.

Terrestrial herb. Miombo woodland. 1400–1790 m.

Distribution: MP, SL—*Gereau 3874*

Sacciolepis typhura (Stapf) Stapf; Clayton & Renvoize, F.T.E.A. (Gramineae 3): 460. 1982.

Terrestrial herb. Stream banks in montane grassland. 2560–2570 m.

Distribution: KP—*Gereau 3169*

Setaria homonyma (Steud.) Chiov.; Clayton & Renvoize, F.T.E.A. (Gramineae 3): 536. 1982.

Terrestrial herb. Riverine forest, grassland. 1200–1350 m.

Distribution: MP—*J. Lovett 4645*

Setaria plicatilis (Hochst.) Hack. ex Engl.; Clayton & Renvoize, F.T.E.A. (Gramineae 3): 538. 1982.

Terrestrial herb. Montane forest. 1750 m.

Distribution: RM—*Abeid 2899*

Setaria pumila (Poir.) Roem. & Schult.; Clayton & Renvoize, F.T.E.A. (Gramineae 3): 530. 1982.

Terrestrial herb. Paths in montane forest, roadsides in miombo woodland, abandoned cultivation. 1200–2110 m.

Distribution: MP, MR, SL—*Gereau 4564*

Setaria sphacelata (Schumach.) M. B. Moss ex Stapf & C. E. Hubb.; Clayton & Renvoize, F.T.E.A. (Gramineae 3): 527. 1982.

Terrestrial herb. Woodland, seasonally inundated grassland, fallow fields. 1270–2300 m.

Distribution: MP, MR—*Gereau 3482*

Sinarundinaria alpina (K. Schum. ex Engl.) C. S. Chao & Renvoize; Lovett et al., 2006: 289.

Bamboo. Montane forest edges, heathland. 2000–2730 m.

Distribution: MR, NL, RM—*Gereau 4535*

Sorghastrum bipennatum (Hack.) Pilg.; Clayton & Renvoize, F.T.E.A. (Gramineae 3): 731. 1982.

Terrestrial herb. Grassland, miombo woodland. 1250–1800 m.

Distribution: MP—*P. Lovett 116*

Sporobolus africanus (Poir.) Robyns & Tournay; Clayton, F.T.E.A. (Gramineae 2): 375. 1974.

Terrestrial herb. Montane grassland, clearings in montane forest, edges of cultivation. 1900–2500 m.

Distribution: KP, NL, SL—*Gereau 4770*

Sporobolus myrianthus Benth.; Clayton, F.T.E.A. (Gramineae 2): 383. 1974.

Terrestrial herb. Riverine forest, woodland, abandoned cultivation. 1200–1600 m.

Distribution: MP—*Gereau 3491*

Sporobolus natalensis (Steud.) T. Durand & Schinz; Clayton, F.T.E.A. (Gramineae 2): 374. 1974.

Terrestrial herb. Montane grassland. 2080–2140 m.

Distribution: SL—*Suleiman 110*

Sporobolus pyramidalis P. Beauv.; Clayton, F.T.E.A. (Gramineae 2): 373. 1974.

Terrestrial herb. Seasonal pond margins in montane grassland, abandoned cultivation. 1250–1840 m.

Distribution: MP, SL—*Gereau 3912*

Sporobolus subtilis Kunth; Clayton, F.T.E.A. (Gramineae 2): 386. 1974.

Terrestrial herb. Montane grassland. 2100 m.

Distribution: RM—*Stolz 1073*

Themeda triandra Forssk.; Clayton & Renvoize, F.T.E.A. (Gramineae 3): 829. 1982.

Terrestrial herb. Montane grassland, miombo woodland, edges of cultivation. 1260–2800 m.

Distribution: KP, SL—*Suleiman 153*

Trachypogon spicatus (L. f.) Kuntze; Clayton & Renvoize, F.T.E.A. (Gramineae 3): 709. 1982.

Terrestrial herb. Montane grassland. 2320 m.

Distribution: SL—*Gereau 3994*

Trichopteryx elegantula (Hook. f.) Stapf; Clayton, F.T.E.A. (Gramineae 2): 410. 1974.

Terrestrial herb. Rocky woodland. 1300–1600 m.

Distribution: MP—*J. Lovett 4380*

Trichopteryx marungensis Chiov.; Clayton, F.T.E.A. (Gramineae 2): 411. 1974.

Terrestrial herb. Montane grassland. 1940 m.

Distribution: SL—*Gereau 4754*

Tripogon major Hook. f.; Phillips, F.T.E.A. (Gramineae 2): 291. 1974.

Terrestrial herb. Rocky woodland. 2100 m.

Distribution: SL—*Gereau 4019*

Tristachya bequaertii De Wild.; Clayton, F.T.E.A. (Gramineae 2): 428. 1974.

Terrestrial herb. Woodland. 1200 m.

Distribution: MP—*J. Lovett 3940*

Urochloa trichopus (Hochst.) Stapf; Clayton & Renvoize, F.T.E.A. (Gramineae 3): 604. 1982.

Terrestrial herb. Roadsides. 474–490 m.

Distribution: SL—*Gereau 3799*

Zonotriche inamoena (K. Schum.) Clayton; Clayton, F.T.E.A. (Gramineae 2): 429. 1974.

Terrestrial herb. Miombo woodland. 1260–1530 m.

Distribution: SL—*Gereau 4385*

POLYGALACEAE

3 genera and 5 species

Muraltia flanaganii Bolus; Paiva, F.T.E.A. (Polygalaceae): 8. 2007.

Suffrutex. Rocky mountain ridges. 2800 m.

Distribution: KP, MR—*Goldblatt 8191*

Polygala exelliana Troupin; Paiva, F.T.E.A. (Polygalaceae): 41. 2007.

Shrub. Montane forest edges, shrubland. 2110–2350 m.

Distribution: MR, RM—*Gereau 4549*

Polygala macrostigma Chodat; Paiva, F.T.E.A. (Polygalaceae): 40. 2007.

Terrestrial herb. Woodland, montane grassland. 1400–1940 m.

Distribution: MP, SL—*Gereau 4685*

Polygala virgata Thunb. var. *decora* (Sond.) Harv.; Paiva, F.T.E.A. (Polygalaceae): 56. 2007.

Shrub. Bamboo forest.

Distribution: KP, MP—*Brenan 8205*

Securidaca longipedunculata Fresen.; Paiva, F.T.E.A. (Polygalaceae): 6. 2007.

Shrub or tree. Miombo woodland. 474–1475

m.

Distribution: MP, NL—*Kayombo 2430*

POLYGONACEAE

4 genera and 8 species

Antigonon leptopus Hook. & Arn.; Graham, F.T.E.A. (Polygonaceae): 38. 1958.

Lianescent shrub. Lakeshore forest edges. 474–490 m.

Distribution: SL—*Gereau 3795*

Fagopyrum snowdenii (Hutch. & Dandy) S. P. Hong; Hong, 1988: 295.

Terrestrial herb. Riverbanks in montane grassland. 2560–2570 m.

Distribution: KP—*Gereau 3164*

Persicaria decipiens (R. Br.) K. L. Wilson; Nogueira et al., Fl. Zamb. 9(3): 9. 2006.

Terrestrial or emergent aquatic herb. Riverine forest. 1210–1500 m.

Distribution: SL—*Gereau 4269*

Persicaria madagascariensis (Meisn.) S. Ortiz & Paiva; Nogueira et al., Fl. Zamb. 9(3): 12. 2006.

Terrestrial herb. Damp soil in woodland. 1300 m.

Distribution: MP—*J. Lovett 3446*

Persicaria nepalensis (Meisn.) H. Gross; Nogueira et al., Fl. Zamb. 9(3): 5. 2006.

Terrestrial herb. Edges of pyrethrum fields. 2200 m.

Distribution: MR—*Gereau 4513*

Rumex abyssinicus Jacq.; Graham, F.T.E.A. (Polygonaceae): 7. 1958.

Terrestrial herb. Montane forest and grassland, seasonal marshland. 1770–2840 m.

Distribution: KP, RM, SL—*Abeid 2876*

Rumex bequaertii De Wild.; Graham, F.T.E.A. (Polygonaceae): 8. 1958.

Terrestrial herb. Rocky montane grassland. 2960 m.

Distribution: KP, MR—*St. Clair-Thompson 820*

Rumex usambarensis (Dammer) Dammer; Graham, F.T.E.A. (Polygonaceae): 7. 1958.

Terrestrial herb. Woodland, montane grassland. 1300–2580 m.

Distribution: KP, MP, RM—*Abeid 2840*

PRIMULACEAE

2 genera and 6 species

Anagallis brevipes P. Taylor; Taylor, F.T.E.A. (Primulaceae): 13. 1958.

Terrestrial herb. Wet soil in montane grassland. 2250 m.

Distribution: RM—*Greenway 8278* (type: EA, K)

Anagallis djalonis A. Chev.; Taylor, F.T.E.A. (Primulaceae): 17. 1958.

Terrestrial herb. Roadsides in montane grassland, riverine forest. 1500–1940 m.

Distribution: RM, SL—*Gereau 4669*

Anagallis hexamera P. Taylor; Taylor, F.T.E.A. (Primulaceae): 15. 1958.

Terrestrial or aquatic herb. Swamp. 2100 m.

Distribution: MP—*Richards 17024*

Anagallis kingaensis Engl.; Taylor, F.T.E.A. (Primulaceae): 11. 1958.

Terrestrial herb. Stream banks in montane grassland. 2100–2700 m.

Distribution: KP, RM—*Goldblatt 8186*

Ardisiandra primuloides Knuth; Tayor, F.T.E.A. (Primulaceae): 2. 1958.

Terrestrial herb. Shaded rocky ground. 1000–2250 m.

Distribution: RM—*Stolz 924* (type: K, UPS), *Geilinger 2365*

Ardisiandra wettsteinii R. Wagner; Taylor, F.T.E.A. (Primulaceae): 4. 1958.

Terrestrial herb. Montane forest, montane grassland and shrubland. 2120–2740 m.

Distribution: KP, MR, RM, SL—*Gereau 4343*

PROTEACEAE

2 genera and 9 species

Faurea rochetiana (A. Rich.) Chiov. ex Pic. Serm.; Marner, F.T.E.A. (Proteaceae): 3. 1993.

Tree or shrub. Woodland, riverine forest. 1200–2500 m.

Distribution: KP, MP, SL—*Kayombo 2399*

Faurea wentzeliana Engl.; Marner, F.T.E.A. (Proteaceae): 6. 1993.

Tree. Woodland, montane grassland. 1520–1590 m.

Distribution: NL, RM—*Bracebridge 23*

Protea gaguedi J. F. Gmel.; Brummitt, F.T.E.A. (Proteaceae): 17. 1993.

Tree or shrub. Woodland, montane grassland. 1300–2140 m.

Distribution: MP, SL—*Kayombo 2402*

Protea heckmanniana Engl. subsp. *heckmanniana*; Brummitt, F.T.E.A. (Proteaceae): 25. 1993.

Suffrutex. Montane grassland. 2120–2700 m.

Distribution: MP, SL—*Gereau 4324*

Protea humifusa Beard; Brummitt, F.T.E.A. (Proteaceae): 26. 1993.

Suffrutex. Montane grassland, rocky ridges. 2010–2800 m.

Distribution: KP, MP, RM—*Napper 1173* (type: EA, K), *Taylor 8523*

Protea linearifolia Engl.; Brummitt, F.T.E.A. (Proteaceae): 26. 1993.

Suffrutex. Rocky mountain slopes. 2600 m.

Distribution: KP—*Brummitt 18136*

Protea rupestris R. E. Fr.; Brummitt, F.T.E.A. (Proteaceae): 17. 1993.

Tree. Miombo woodland. 1580 m.

Distribution: SL—*Gereau 4809*

Protea welwitschii Engl.; Brummitt, F.T.E.A. (Proteaceae): 19. 1993.

Shrub. Montane grassland, miombo woodland, secondary shrubland. 1760–2230 m.

Distribution: NL, SL—*Gereau 3906*

Protea wentzeliana Engl.; Brummitt, F.T.E.A. (Proteaceae): 22. 1993.

Shrub. Tree clumps in montane grassland. 2360 m.

Distribution: MR—*Gereau 3738*

RANUNCULACEAE

4 genera and 13 species

Clematis chrysocarpa Welw. ex Oliv. subsp. *bijuga* Brummitt; Brummitt, 2000: 102.

Terrestrial herb. Miombo woodland. 1400–1530 m.

Distribution: MP, SL—*Gereau 4363*

Clematis hirsuta Guill. & Perr.; Milne-Redhead & Turrill, F.T.E.A. (Ranunculaceae): 6. 1952.

Liana, lianescent shrub, or herbaceous climber. Woodland, riverine forest, montane forest, bamboo thicket. 840–2090 m.

Distribution: MP, MR, NL, RM, SL—*Gereau 4600*

Clematis simensis Fresen.; Milne-Redhead & Turrill, F.T.E.A. (Ranunculaceae): 2. 1952.

Herbaceous climber. Montane forest and grassland. 2100–2400 m.

Distribution: MP, MR—*Mwasumbi 16316*

Clematis uhehensis Engl.; Brummitt, 2000: 105.

Terrestrial herb. Montane grassland, rocky hills, woodland. 1200–2800 m.

Distribution: KP, MP, MR, RM, SL—*Suleiman 16*

Clematis villosa DC. subsp. *oliveri* (Hutch.) Brummitt; Brummitt, 2000: 104.

Woody-based terrestrial herb. Montane grassland, riverine forest. 1100–1940 m.

Distribution: MP, RM, SL—*Gereau 4773*

Clematis welwitschii Hiern ex Kuntze; Milne-Redhead & Turrill, F.T.E.A. (Ranunculaceae): 3. 1952.

Herbaceous climber. Riverine forest. 1210 m.

Distribution: SL—*Gereau 4288*

Delphinium leroyi Franch. ex Huth; Milne-Redhead & Turrill, F.T.E.A. (Ranunculaceae): 20. 1952.

Terrestrial herb. Montane grassland. 1940–2590 m.

Distribution: KP, MP, RM, SL—*Suleiman 17*

Ranunculus multifidus Forssk.; Milne-Redhead & Turrill, F.T.E.A. (Ranunculaceae): 19. 1952.

Terrestrial and emergent aquatic herb. Riverbanks and ponds in montane grassland, streambeds in montane forest, woodland, riverine forest. 1400–2550 m.

Distribution: KP, MP, MR, RM, SL—*Gereau 4625*

Ranunculus oreophytus Delile; Milne-Redhead & Turrill, F.T.E.A. (Ranunculaceae): 14. 1952.

Terrestrial herb. Stream banks in montane grassland. 2500 m.

Distribution: KP—*Goldblatt 8187*

Ranunculus raeae Exell; Milne-Redhead & Turrill, F.T.E.A. (Ranunculaceae): 16. 1952.

Terrestrial herb. Montane grassland. 2400–2600 m.

Distribution: KP—*Suleiman 47*

Ranunculus volkensii Engl.; Milne-Redhead & Turrill, F.T.E.A. (Ranunculaceae): 14. 1952.

Terrestrial or aquatic herb. Stream banks and swamps in montane grassland. 2500–2550 m.

Distribution: KP—*Goldblatt 8248*

Thalictrum rhynchocarpum Quart.-Dill. & A. Rich.; Milne-Redhead & Turrill, F.T.E.A. (Ranunculaceae): 11. 1952.

Terrestrial herb. Montane grassland, riverbank thickets, montane forest, scree slopes, lava flows. 1980–2560 m.

Distribution: KP, MR, RM, SL—*Gereau 3720*

Thalictrum zernyi Ulbr.; Milne-Redhead & Turrill, F.T.E.A. (Ranunculaceae): 9. 1952.

Terrestrial herb. Riverbanks in montane forest and grassland, forest edges in montane grassland. 2070–2590 m.

Distribution: KP, NL, SL—*Gereau 4136*

RHAMNACEAE

4 genera and 6 species

Phylica emirnensis (Tul.) Pillans; Johnston, F.T.E.A. (Rhamnaceae): 9. 1972.

Suffrutex. Rocky mountain ridges. 2600–2800 m.

Distribution: KP—*Goldblatt 8194*

Rhamnus prinoides L'Hér.; Johnston, F.T.E.A. (Rhamnaceae): 18. 1972.

Shrub or tree. Montane forest, bamboo forest. 1590–2400 m.

Distribution: NL, RM—*Abeid 2812*

Rhamnus staddo A. Rich.; Johnston, F.T.E.A. (Rhamnaceae): 18. 1972.

Shrub. Tree clumps in montane grassland, bamboo forest. 2360–2630 m.

Distribution: KP, MR—*Gereau 3719*

Scutia myrtina (Burm. f.) Kurz; Johnston, F.T.E.A. (Rhamnaceae): 21. 1972.

Liana. Disturbed montane forest.

Distribution: RM—*Ndangalasi 959*

Ziziphus abyssinica Hochst. ex A. Rich.; Johnston, F.T.E.A. (Rhamnaceae): 27. 1972.

Tree. Woodland. 1200–1400 m.

Distribution: MP—*Kayombo 2433*

Ziziphus mucronata Willd. subsp. *mucronata*; Johnston, F.T.E.A. (Rhamnaceae): 26. 1972.

Tree. Riverine forest. 840 m.

Distribution: SL—*Kibure 1457*

RHIZOPHORACEAE

1 genus and 2 species

Cassipourea gummiflua Tul.; Lewis, F.T.E.A. (Rhizophoraceae): 15. 1956.

Tree or shrub. Montane forest. 1750–1965 m.

Distribution: RM—*Abeid 2819*

Cassipourea malosana (Baker) Alston; Lewis, F.T.E.A. (Rhizophoraceae): 11. 1956.

Tree. Montane forest. 1885–2320 m.

Distribution: NL, RM—*Kibure 1363*

ROSACEAE

6 genera and 11 species

Alchemilla kiwuensis Engl.; Graham, F.T.E.A. (Rosaceae): 14. 1960.

Terrestrial herb. Trails and stream banks in montane forest, roadsides. 1790–2130 m.

Distribution: MR, NL, RM—*Gereau 4163*

Cliffortia nitidula R. E. Fr. & T. C. E. Fr.; Graham, F.T.E.A. (Rosaceae): 42. 1960.

Shrub. Montane grassland, riversides. 2550–2740 m.

Distribution: KP, RM—*Prins 200*

Fragaria vesca L.; Graham, F.T.E.A. (Rosaceae): 20. 1960.

Terrestrial herb. Heathland. 2950 m.

Distribution: KP—*Richards 6589*

Hagenia abyssinica J. F. Gmel.; Graham, F.T.E.A. (Rosaceae): 43. 1960.

Tree. Montane forest, mostly at edges. 1920–2230 m.

Distribution: MR, RM, SL—*Gereau 3570*

Prunus africana (Hook. f.) Kalkman; Lovett et al. 2006: 207.

Tree. Riverine forest, montane forest. 1600–2700 m.

Distribution: MP, NL, RM, SL—*Harder 1507*

Rubus apetalus Poir.; Graham, F.T.E.A. (Rosaceae): 39. 1960.

Scrambling shrub. Montane forest. 2080–2300 m.

Distribution: MP, RM, SL—*Suleiman 146*

Rubus ellipticus Sm.; Graham, F.T.E.A. (Rosaceae): 25. 1960.

Scrambling shrub. Montane forest. 1780–2250 m.

Distribution: RM—*Abeid 2919*

Rubus kirungensis Engl.; Graham, F.T.E.A. (Rosaceae): 36. 1960.

Scrambling shrub.

Distribution: RM—*MacInnes 408*

Rubus niveus Thunb.; Graham, F.T.E.A. (Rosaceae): 40. 1960.

Scrambling shrub. Edges of cultivation. 1250 m.

Distribution; MP—*J. Lovett 3843*

Rubus pinnatus Willd.; Graham, F.T.E.A. (Rosaceae): 37. 1960.

Scrambling shrub. Montane grassland, riverine forest, disturbed montane forest. 1780–1940 m.

Distribution: MP, RM, SL—*J. Lovett 4876*

Rubus porotoensis R. A. Graham; Graham, F.T.E.A. (Rosaceae): 35. 1960.

Scrambling shrub. Montane forest edges. 2100–2340 m.

Distribution: MR—*Richards 9700* (type: K), *Michelmore 965*

RUBIACEAE

49 genera and 119 species

Afrocanthium parasiebenlistii (Bridson) Lantz vel sp. aff.; Lantz & Bremer, 2004: 278.

Tree. Montane forest edges, riverine forest. 1320–2230 m.

Distribution: MP, SL—*Gereau 3580*

Agathisanthemum quadricostatum Bremek.; Verdcourt, F.T.E.A. (Rubiaceae 1): 256. 1976.

Terrestrial herb. Miombo woodland. 1530 m.

Distribution: SL—*Gereau 4376*

Anthospermum herbaceum L. f.; Verdcourt, F.T.E.A. (Rubiaceae 1): 325. 1976.

Terrestrial herb. Montane grassland.

Distribution: RM—*Ndangalasi 994*

Anthospermum rosmarinus K. Schum.; Verdcourt, F.T.E.A. (Rubiaceae 1): 329. 1976.

Shrub. Montane grassland. 1940–2150 m.

Distribution: NL, SL—*Gereau 4660A*

Anthospermum usambarense K. Schum.; Verdcourt, F.T.E.A. (Rubiaceae 1): 331. 1976.

Shrub. Montane grassland, roadsides. 1940–2770 m.

Distribution: KP, MP, NL, RM, SL—*Abeid 2870*

Anthospermum whyteanum Britten; Verdcourt, F.T.E.A. (Rubiaceae 1): 331. 1976.

Subshrub. Montane grassland. 2440 m.

Distribution: SL—*Suleiman 185*

Catunaregam spinosa (Thunb.) Tirveng. subsp. *spinosa*; Verdcourt, F.T.E.A. (Rubiaceae 2): 499. 1988.

Shrub. Lakeshore forest. 474 m.

Distribution: SL—*Gereau 3663*

Catunaregam spinosa (Thunb.) Tirveng. subsp. *taylorii* (S. Moore) Verdc.; Verdcourt, F.T.E.A. (Rubiaceae 2): 500. 1988.

Shrub or tree. Woodland. 1250–1400 m.

Distribution: MP—*Kayombo 966*

Cephalanthus natalensis Oliv.; Ridsdale, 1976: 183.

Liana. Montane forest. 1900–2040 m.

Distribution: NL, SL—*Gereau 4733*

Note: New to East Africa, from South Africa and Malawi.

Chamaepentas nobilis (S. Moore) Kårehed & B. Bremer; Kårehed & Bremer, 2007: 1075.

Terrestrial herb. Woodland. 1200–1250 m.

Distribution: MP—*J. Lovett 4227*

Chassalia discolor K. Schum. subsp. *discolor*; Verdcourt, F.T.E.A. (Rubiaceae 1): 127. 1976.

Shrub. Submontane forest. 1630 m.

Distribution: RM—*WCS SHCP–142*

Chassalia parvifolia K. Schum.; Verdcourt, F.T.E.A. (Rubiaceae 1): 130. 1976.

Shrub. Montane forest, riverine forest. 1560–2260 m.

Distribution: MR, RM, SL—*Gereau 4710*

Chassalia subochreata (De Wild.) Robyns; Verdcourt, F.T.E.A. (Rubiaceae 1): 131. 1976.

Tree. Montane forest. 1950 m.

Distribution: RM—*Bracebridge 88*

Coffea canephora Pierre ex A. Froehner; Bridson, F.T.E.A. (Rubiaceae 2): 710. 1988.

Shrub. Riverine forest. 1300 m.

Distribution: MP—*Kayombo 2353*

Coffea mufindiensis Hutch. ex Bridson subsp. *mufindiensis*; Bridson, F.T.E.A. (Rubiaceae 2): 720. 1988.

Shrub. Montane forest, riverine forest. 2000–2080 m.

Distribution: MP, MR, SL—*Gereau 4735*

Cremaspora triflora (Thonn.) K. Schum. subsp. *confluens* (K. Schum.) Verdc.; Verdcourt, F.T.E.A. (Rubiaceae 2): 735. 1988.

Shrub. Riverine forest. 474–910 m.

Distribution: SL—*Mwasumbi 16600*

Cremaspora triflora (Thonn.) K. Schum. subsp. *triflora*; Verdcourt, F.T.E.A. (Rubiaceae 2): 733. 1988.

Shrub. Stream banks in miombo woodland. 980 m.

Distribution: SL—*Harder 1529*

Crossopteryx febrifuga (Afzel. ex G. Don) Benth.; Verdcourt, F.T.E.A. (Rubiaceae 2): 457. 1988.

Tree or shrub. Woodland, lakeshore forest. 480–1400 m.

Distribution: MP, SL—*Gereau 3646*

Diodia sarmentosa Sw.; Verdcourt, F.T.E.A. (Rubiaceae 1): 336. 1976.

Terrestrial herb. Woodland/montane forest transition. 2140 m.

Distribution: SL—*Gereau 4012*

Dolichopentas decora (S. Moore) Kårehed & B. Bremer var. *decora* Moore; Kårehed & Bremer, 2007: 1076.

Terrestrial herb. Woodland, montane grassland. 1700–2520 m.

Distribution: KP, MP, MR, SL—*Gereau 4434*

Dolichopentas lindenioides (S. Moore) Kårehed & B. Bremer; Kårehed & Bremer, 2007: 1076.

Terrestrial herb. Rocky mountain slopes, montane grassland. 2400–2820 m.

Distribution: KP, MP, MR, NL, RM—*Suleiman 64*

Fadogia ancylantha Hiern; Verdcourt, F.T.E.A. (Rubiaceae 3): 799. 1991.

Suffrutex or shrub. Woodland, abandoned cultivation. 1200–1720 m.

Distribution: MP, SL—*Kayombo 2482*

Fadogia elskensii De Wild.; Verdcourt, F.T.E.A. (Rubiaceae 3): 790. 1991.

Suffrutex. Woodland. 2000 m.

Distribution: SL—*Gereau 4066*

Fadogia stenophylla Hiern subsp. *odorata* (K. Krause) Verdc.; Verdcourt, F.T.E.A. (Rubiaceae 3): 794. 1991.

Terrestrial herb. Montane grassland, woodland. 1200–2250 m.

Distribution: MP, SL—*Gereau 5135*

Fadogia triphylla Baker var. *giorgii* (De Wild.) Verdc.; Verdcourt, F.T.E.A. (Rubiaceae 3): 797. 1991.

Shrub. Miombo woodland. 1500 m.

Distribution: MP—*Gereau 6996*

Fadogia triphylla Baker var. *triphylla*; Verdcourt, F.T.E.A. (Rubiaceae 3): 796. 1991.

Terrestrial herb or suffrutex. Woodland, roadsides in village. 1200–1400 m.

Distribution: MP, SL—*Gereau 3567*

Galiniera saxifraga (Hochst.) Bridson; Bridson, F.T.E.A. (Rubiaceae 2): 696. 1988.

Shrub or tree. Montane forest, riverine forest. 1560–2300 m.

Distribution: MR, NL, SL, RM—*Suleiman 75*

Galium bussei K. Schum. & K. Krause; Verdcourt, F.T.E.A. (Rubiaceae 1): 399. 1976.

Terrestrial herb. Montane grassland, rocky mountain ridges, woodland, edges of cultivation. 1250–2700 m.

Distribution: KP, MP, SL—*Suleiman 184*

Galium chloroionanthum K. Schum.; Verdcourt, F.T.E.A. (Rubiaceae 1): 388. 1976.

Terrestrial herb, often climbing. Montane forest edges. 2700 m.

Distribution: MR, RM—*Goetze 1162* (type: BM, K), *Aleljung 516*

Galium ossirwaense K. Krause; Verdcourt, F.T.E.A. (Rubiaceae 1): 400. 1976.

Terrestrial herb. Montane grassland. 2260–2700 m.

Distribution: KP, MR, NL, RM—*Suleiman 62*

Galium scabrellum K. Schum.; Verdcourt, F.T.E.A. (Rubiaceae 1): 401. 1976.

Terrestrial herb, often climbing. Montane grassland, woodland, scree slopes. 1280–2135 m.

Distribution: MP, RM, SL—*Kayombo 2371*

Galium stenophyllum Baker; Verdcourt, F.T.E.A. (Rubiaceae 1): 397. 1976.

Terrestrial herb. Riverine forest, miombo woodland. 1300–1800 m.

Distribution: MP—*J. Lovett 3591*

Galium tanganyikense Ehrend. & Verdc.; Verdcourt, F.T.E.A. (Rubiaceae 1): 397. 1976.

Terrestrial herb. Montane grassland. 2490–2590 m.

Distribution: KP, NL—*Stolz 2282* (type: BM, K, MO), *Taylor 8521*

Gardenia subacaulis Stapf & Hutch.; Verdcourt, F.T.E.A. (Rubiaceae 2): 510. 1988.

Suffrutex. Woodland. 1250 m.

Distribution: MP—*J. Lovett 3561*

Gardenia ternifolia Schumach. & Thonn. subsp. *jovis-tonantis* (Welw.) Verdc.; Verdcourt, F.T.E.A. (Rubiaceae 2): 509. 1988.

Tree or shrub. Woodland. 1200–1400 m.

Distribution: MP—*Kayombo 2434*

Hedythyrsus thamnoideus (K. Schum.) Bremek.; Verdcourt, F.T.E.A. (Rubiaceae 1): 250. 1976.

Shrub or perennial herb. Montane grassland, rocky mountain ridges. 2150–2800 m.

Distribution: KP, RM, SL—*Gereau 5134*

Heinsenia diervilleoides K. Schum. subsp. *diervilleoides*; Verdcourt, F.T.E.A. (Rubiaceae 2): 732. 1988.

Tree. Montane forest. 1780–1980 m.

Distribution: RM—*Kibure 1371*

Hymenodictyon floribundum (Hochst. & Steud.) B. L. Rob.; Verdcourt, F.T.E.A. (Rubiaceae 2): 452. 1988.

Tree or lianescent shrub. Woodland, montane forest. 1200–1920 m.

Distribution: MP, SL—*Kayombo 2478*

Ixora narcissodora K. Schum.; Bridson, F.T.E.A. (Rubiaceae 2): 616. 1988.

Shrub. Riverine forest. 980–1210 m.

Distribution: SL—*Gereau 4276*

Ixora scheffleri K. Schum. & K. Krause subsp. *scheffleri*; Bridson, F.T.E.A. (Rubiaceae 2): 613. 1988.

Tree or shrub. Montane forest. 1750–2040 m.

Distribution: MR, NL, RM, SL—*Abeid 2776*

Keetia gueinzii (Sond.) Bridson; Bridson, F.T.E.A. (Rubiaceae 3): 911. 1991.

Shrub or liana. Riverine forest, montane forest, woodland. 1100–2300 m.

Distribution: KP, MP, MR, NL, SL—*Mwangoka 659*

Keetia venosa (Oliv.) Bridson; Bridson, F.T.E.A. (Rubiaceae 3): 914. 1991.

Lianescent shrub. Riverine forest. 1210 m.

Distribution: SL—*Gereau 4286*

Kohautia coccinea Royle; Verdcourt, F.T.E.A. (Rubiaceae 1): 235. 1976.

Terrestrial herb. Woodland, abandoned cultivation. 1250–2000 m.

Distribution: MP—*J. Lovett 4605*

Kohautia longifolia Klotzsch var. *longifolia*; Verdcourt, F.T.E.A. (Rubiaceae 1): 233. 1976.

Terrestrial herb. Grassland. 1550 m.

Distribution: RM—*Aleljung 352*

Mitracarpus hirtus (L.) DC.; Verdcourt, Fl. Zamb. 5(1): 188. 1989.

Terrestrial herb. Disturbed riverine forest. 1150 m.

Distribution: MP—*Kayombo 2451*

Mitragyna rubrostipulata (K. Schum.) Havil.; Lovett et al., 2006: 217.

Tree. Montane forest. 1900 m.

Distribution: RM—*Kibure 1362*

Multidentia crassa (Hiern) Bridson & Verdc. var. *crassa*; Bridson, F.T.E.A. (Rubiaceae 3): 847. 1991.

Tree or shrub. Miombo woodland, riverine forest. 1200–1920 m.

Distribution: MP, RM—*Kayombo 1093*

Multidentia fanshawei (Tennant) Bridson vel sp. aff.; Bridson, F.T.E.A. (Rubiaceae 3): 843. 1991.

Tree. Montane forest. 1860 m.

Distribution: RM—*Bracebridge 81*

Mussaenda arcuata Lam ex Poir.; Bridson, F.T.E.A. (Rubiaceae 2): 461. 1988.

Liana. Miombo woodland, secondary shrubland, wooded grassland, riverine forest, montane forest. 950–1760 m.

Distribution: NL, RM, SL—*Gereau 3849*

Oldenlandia corymbosa L. var. *caespitosa* (Benth.) Verdc.; Verdcourt, F.T.E.A. (Rubiaceae 1): 310. 1976.

Terrestrial herb. Sandy woodland edges. 1920 m.

Distribution: MR—*W. H. Lewis 6088*

Note: This collection is somewhat intermediate with *O. capensis* L. f. var. *capensis*.

Oldenlandia echinulosa K. Schum. var. *echinulosa*; Verdcourt, Fl. Zamb. 5(1): 130. 1989.

Terrestrial herb. Cultivated fields. 1700–2000 m.

Distribution: MP, RM—*Mwasumbi 16234*

Oldenlandia echinulosa K. Schum. var. *pellucida* (Hiern) Verdc.; Verdcourt, Fl. Zamb. 5(1): 130. 1989.

Terrestrial herb. Cultivated fields, riverine forest. 1600–1800 m.

Distribution: MP, MR—*P. Lovett 217*

Oldenlandia friesiorum Bremek.; Verdcourt, F.T.E.A. (Rubiaceae 1): 282. 1976.

Terrestrial herb. Montane grassland. 1950–2360 m.

Distribution: MR, NL, RM—*Gereau 3715*

Oldenlandia herbacea (L.) Roxb. var. *herbacea*; Verdcourt, F.T.E.A. (Rubiaceae 1): 305. 1976.

Terrestrial herb. Riverine bushland. 1600 m.

Distribution: RM—*Aleljung 321*

Oldenlandia herbacea (L.) Roxb. var. *holstii* (K. Schum.) Bremek.; Verdcourt, F.T.E.A. (Rubiaceae 1): 306. 1976.

Terrestrial herb. 1980 m.

Distribution: MP—*Leedal 4259*

Oldenlandia rosulata K. Schum.; Verdcourt, F.T.E.A. (Rubiaceae 1): 290. 1976.

Terrestrial herb. Montane grassland, miombo woodland. 1800–1940 m.

Distribution: MP, SL—*Gereau 4679*

Oldenlandia rupicola (Sond.) Kuntze var. *rupicola*; Verdcourt, F.T.E.A. (Rubiaceae 1): 284. 1976.

Prostrate herb or subshrub. Montane heathland. 2730 m.

Distribution: RM—*Abeid 2989*

Otiophora caerulea (Hiern) Bullock; Verdcourt, F.T.E.A. (Rubiaceae 1): 321. 1976.

Terrestrial herb. Woodland, montane grassland. 1700–2500 m.

Distribution: MP—*J. Lovett 4170*

Otiophora stolzii (Verdc.) Verdc.; Verdcourt, F.T.E.A. (Rubiaceae 1): 320. 1976.

Terrestrial herb. Montane grassland. 2080–2520 m.

Distribution: KP, SL—*Stolz 2467* (type: BM, BOL, BR, EA, K, MO), *Suleiman 85*

Otiophora villicaulis Mildbr. var. *villicaulis*; Verdcourt, F.T.E.A. (Rubiaceae 1): 322. 1976.

Terrestrial herb. Miombo woodland, montane grassland. 1530–2230 m.

Distribution: SL—*Gereau 4431*

Oxyanthus speciosus DC. subsp. *stenocarpus* (K. Schum.) Bridson; Bridson, F.T.E.A. (Rubiaceae 2): 529. 1988.

Tree or shrub. Montane forest, riverine forest. 1760–2200 m.

Distribution: MP, MR, NL, RM, SL—*Gereau 4727*

Pauridiantha paucinervis (Hiern) Bremek.; Verdcourt, F.T.E.A. (Rubiaceae 1): 153. 1976.

Shrub. Montane forest. 1560–2320 m.

Distribution: KP, RM, SL—*Gereau 4200*

Pauridiantha udzungwaensis Ntore & Dessein; Ntore et al., 2003: 67.

Lianescent shrub or small tree. Montane forest. 1910–2700 m.

Distribution: RM, SL—*Gereau 4702*

Pavetta cataractarum S. Moore; Bridson, F.T.E.A. (Rubiaceae 2): 647. 1988.

Shrub. Riverine forest. 1160 m.

Distribution: SL—*Gereau 3702*

Pavetta comostyla S. Moore subsp. *nyassica* (Bremek.) Bridson; Bridson, F.T.E.A. (Rubiaceae 2): 656. 1988.

Shrub or tree. Montane forest. 1560–2110 m.

Distribution: MR, RM—*Suleiman 73*

Pavetta crassipes K. Schum.; Bridson, F.T.E.A. (Rubiaceae 2): 670. 1988.

Tree or shrub. Woodland. 1200–1410 m.

Distribution: MP, SL—*Kayombo 1012*

Pavetta gardeniifolia Hochst. ex A. Rich. var. *gardeniifolia*; Bridson, F.T.E.A. (Rubiaceae 2): 677. 1988.

Shrub. Riverine forest. 1200–1500 m.

Distribution: MP, SL—*Gereau 4399*

Pavetta gardeniifolia A. Rich. var. *subtomentosa* K. Schum.; Bridson, F.T.E.A. (Rubiaceae 2): 678. 1988.

Shrub. Riverine forest. 1210 m.

Distribution: SL—*Gereau 4259*

Pavetta hymenophylla Bremek.; Bridson, F.T.E.A. (Rubiaceae 2): 632. 1988.

Tree. Montane forest. 1750–1780 m.

Distribution: RM—*Abeid 2770*

Pavetta kyimbilensis Bremek. var. *iringensis* (Bremek.) Bridson; Bridson, F.T.E.A. (Rubiaceae 2): 667. 1988.

Shrub or tree. Montane forest. 1780–2230 m.

Distribution: MR, RM, SL—*Gereau 4586*

Pavetta kyimbilensis Bremek. var. *kyimbilensis*; Bridson, F.T.E.A. (Rubiaceae 2): 667. 1988.

Shrub. Montane forest. 2100–2620 m.

Distribution: MR, RM—*Stolz 2332* (type: BM, BR, EA, K, MO, P); *Mwasumbi 16215*

Pavetta lasiobractea K. Schum.; Bridson, F.T.E.A. (Rubiaceae 2): 667. 1988.

Tree or shrub. Montane forest. 1800–2040 m.

Distribution: MR, NL, RM, SL—*Harder 1503*

Pavetta schumanniana F. Hoffm. ex K. Schum.; Bridson, F.T.E.A. (Rubiaceae 2): 672. 1988.

Tree or shrub. Woodland, edges of cultivation. 1200–1530 m.

Distribution: MP, SL—*Kayombo 1051*

Pavetta subumbellata Bremek. var. *subumbellata*; Bridson, F.T.E.A. (Rubiaceae 2): 663. 1988.

Shrub. Montane forest. 2000–2320 m.

Distribution: MP, MR, SL—*Gereau 4575*

Pentanisia monticola (K. Krause) Verdc.; Verdcourt, F.T.E.A. (Rubiaceae 1): 170. 1976.

Terrestrial herb. Miombo woodland, riverine forest. 1300–1800 m.

Distribution: MP—*Kayombo 2400*

Pentanisia schweinfurthii Hiern; Verdcourt, F.T.E.A. (Rubiaceae 1): 168. 1976.

Terrestrial herb. Montane grassland, woodland. 1300–2000 m.

Distribution: MP, SL—*Gereau 3867*

Pentas purpurea Oliv. subsp. *purpurea*; Verdcourt, F.T.E.A. (Rubiaceae 1): 207. 1976.

Terrestrial herb. Woodland, grassland, rocky ridges. 1700–2660 m.

Distribution: KP, MP, MR, RM, SL—*Abeid 2843*

Phyllopentas schimperiana (Vatke) Kårehed & B. Bremer; Kårehed & Bremer, 2007: 1076.

Shrub or terrestrial herb. Montane grassland, montane forest patches. 2080–2525 m.

Distribution: RM, SL—*Gereau 3985*

Phyllopentas schumanniana (K. Krause) Kårehed & B. Bremer; Kårehed & Bremer, 2007: 1076.

Shrub, sometimes lianescent. Montane forest, woodland/forest transition. 1900–2260 m.

Distribution: NL, SL—*Gereau 4711*

Phyllopentas ulugurica (Verdc.) Kårehed & B. Bremer; Kårehed & Bremer, 2007: 1076.

Terrestrial herb. Grassland, montane forest. 2000–2100 m.

Distribution: MP—*J. Lovett 3499*

Polysphaeria dischistocalyx Brenan; Verdcourt, F.T.E.A. (Rubiaceae 2): 578. 1988.

Shrub. Riverine forest. 580–1210 m.

Distribution: SL—*Gereau 4257*

Psychotria capensis (Eckl.) Vatke; Verdcourt, Fl. Zamb. 5(1): 13. 1989.

Shrub or tree. Riverine forest. 474–1350 m.

Distribution: MP, SL—*Gereau 4265*

Psychotria ealaensis De Wild.; Verdcourt, F.T.E.A. (Rubiaceae 1): 57. 1976.

Liana. Montane forest. 1910–2040 m.

Distribution: NL, SL—*Gereau 4208*

Psychotria eminiana (Kuntze) E. M. A. Petit var. *eminiana*; Verdcourt & Bridson, F.T.E.A. (Rubiaceae 3): 924. 1991.

Shrub or subshrub. Woodland, edges of cultivation. 1160–1600 m.

Distribution: MP, SL—*Kayombo 1007*

Psychotria goetzei (K. Schum.) E. M. A. Petit; Verdcourt, F.T.E.A. (Rubiaceae 1): 44. 1976.

Shrub. Montane forest. 2200 m.

Distribution: MR—*Gereau 4606*

Psychotria heterosticta E. M. A. Petit var. *plurinervata* E. M. A. Petit; Verdcourt, F.T.E.A. (Rubiaceae 1): 102. 1976.

Shrub. Montane forest. 1960–1980 m.

Distribution: MR—*Mwasumbi 16346*

Psychotria kirkii Hiern var. *diversinodula* Verdc.; Verdcourt, F.T.E.A. (Rubiaceae 1): 98. 1976.

Subshrub. Montane forest. 1960–1980 m.

Distribution: MR—*Mwasumbi 16350*

Psychotria kirkii Hiern var. *nairobiensis* (Bremek.) Verdc.; Verdcourt, F.T.E.A. (Rubiaceae 1): 94. 1976.

Shrub. Lakeshore forest. 480 m.

Distribution: SL—*Gereau 3661*

Psychotria kirkii Hiern var. *swynnertonii* (Bremek.) Verdc.; Verdcourt, F.T.E.A. (Rubiaceae 1): 96. 1976.

Suffrutex. Riverine forest. 1350 m.

Distribution: MP—*J. Lovett 4664*

Psychotria linearisepala E. M. A. Petit var. *linearisepala*; Verdcourt, F.T.E.A. (Rubiaceae 1): 85. 1976.

Shrub. Riverine forest, miombo woodland. 1200–1530 m.

Distribution: MP, SL—*Gereau 4263*

Psychotria mahonii C. H. Wright var. *puberula* (E. M. A. Petit) Verdc.; Verdcourt, F.T.E.A. (Rubiaceae 1): 60. 1976.

Shrub or tree. Montane forest. 1690–2400 m.

Distribution: MR, RM, SL—*Suleiman 77*

Psychotria peduncularis (Salisb.) Steyerm. var. *nyassana* (K. Krause) Verdc.; Verdcourt, F.T.E.A. (Rubiaceae 1): 74. 1976.

Subshrub. Riverine forest. 980 m.

Distribution: SL—*Gereau 5165*

Psychotria schliebenii E. M. A. Petit; Verdcourt, F.T.E.A. (Rubiaceae 1): 80. 1976.

Lianescent shrub. Riverine forest. 1560–1710 m.

Distribution: RM—*Mwasumbi 16282*

Psychotria zombamontana (Kuntze) E. M. A. Petit; Verdcourt, Fl. Zamb. 5(1): 16. 1989.

Shrub or tree. Montane forest, bamboo forest, riverine forest, scree slopes, shrubland. 1580–2600 m.

Distribution: MP, MR, NL, RM, SL—*Abeid 2923*

Psydrax livida (Hiern) Bridson; Bridson, F.T.E.A. (Rubiaceae 3): 898. 1991.

Shrub. Riverine forest, woodland. 1200–1210 m.

Distribution: MP, SL—*Gereau 4287*

Psydrax parviflora (Afzel.) Bridson subsp. *rubrocostata* (Robyns) Bridson; Bridson, F.T.E.A. (Rubiaceae 3): 898. 1991.

Tree. Montane forest. 1725–2040 m.

Distribution: RM, SL—*Gereau 4706*

Psydrax whitei Bridson; Bridson, F.T.E.A. (Rubiaceae 3): 899. 1991.

Tree. Montane forest edges. 2230–2380 m.

Distribution: KP, SL—*Gereau 3575*

Pygmaeothamnus zeyheri (Sond.) Robyns var. *rogersii* Robyns; Verdcourt, F.T.E.A. (Rubiaceae 3): 778. 1991.

Suffrutex. Montane grassland. 2230 m.

Distribution: SL—*Gereau 3594*

Pyrostria lobulata Bridson; Bridson, F.T.E.A. (Rubiaceae 3): 887. 1991.

Shrub. Riverine forest. 980 m.

Distribution: SL—*Harder 1525*

Rhodopentas bussei (K. Krause) Kårehed & B. Bremer; Kårehed & Bremer, 2007: 1076.

Shrub or subshrub. Woodland, edges of cultivation, riverine forest. 1160–1250 m.

Distribution: MP, SL—*Gereau 3701*

Rhodopentas parvifolia (Hiern) Kårehed & B. Bremer; Kårehed & Bremer, 2007: 1076.

Lianescent shrub. Riverine forest. 1210 m.

Distribution: SL—*Gereau 4274*

Rothmannia fischeri (K. Schum.) Bullock subsp. *fischeri*; Bridson, F.T.E.A. (Rubiaceae 2): 513. 1988.

Tree. Riverine forest. 910–1320 m.

Distribution: MP, SL—*Gereau 6123*

Rubia cordifolia L. subsp. *conotricha* (Gand.) Verdc.; Verdcourt, F.T.E.A. (Rubiaceae 1): 381. 1976.

Herbaceous climber. Montane forest, tree clumps in montane grassland, riverine forest, woodland. 1150–2360 m.

Distribution: KP, MP, MR, NL, SL—*Kayombo 2443*

Rutidea fuscescens Hiern subsp. *fuscescens*; Bridson, F.T.E.A. (Rubiaceae 2): 610. 1988.

Liana. Riverine forest. 980 m.

Distribution: SL—*Gereau 5167*

Rutidea orientalis Bridson; Bridson, F.T.E.A. (Rubiaceae 2): 606. 1988.

Lianescent shrub or liana. Montane forest, riverine forest. 1560–1900 m.

Distribution: NL, RM—*Gereau 4182*

Rytigynia adenodonta (K. Schum.) Robyns var. *adenodonta*; Verdcourt, F.T.E.A. (Rubiaceae 3): 819. 1991.

Shrub. Montane forest. 2200–2300 m.

Distribution: MP, MR—*Mwasumbi 16541*

Rytigynia adenodonta (K. Schum.) Robyns var. *reticulata* (Robyns) Verdc.; Verdcourt, F.T.E.A. (Rubiaceae 3): 819. 1991.

Shrub or tree. Riverine forest. 1780 m.

Distribution: MP, RM—*J. Lovett 4873*

Rytigynia bugoyensis (K. Krause) Verdc. subsp. *glabriflora* Verdc.; Verdcourt, F.T.E.A. (Rubiaceae 3): 835. 1991.

Shrub. Montane forest. 1780 m.

Distribution: RM—*Abeid 2766*

Rytigynia flavida Robyns; Verdcourt, F.T.E.A. (Rubiaceae 3): 826. 1991.

Tree or shrub. Montane forest. 1860–2260 m.

Distribution: MR, NL—*Suleiman 72*

Rytigynia cf. *flavida* Robyns; Verdcourt, F.T.E.A. (Rubiaceae 3): 826. 1991.

Tree. Montane forest. 2040 m.

Distribution: SL—*Gereau 4696*

Rytigynia monantha (K. Schum.) Robyns var. *lusakati* Verdc.; Verdcourt, F.T.E.A. (Rubiaceae 3): 826. 1991.

Shrub. Riverine forest. 1210 m.

Distribution: SL—*Gereau 4246*

Rytigynia monantha (K. Schum.) Robyns var. *monantha*; Verdcourt, F.T.E.A. (Rubiaceae 3): 826. 1991.

Shrub. Woodland. 1530–1970 m.

Distribution: SL—*Gereau 4023*

Rytigynia obscura Robyns; Verdcourt, F.T.E.A. (Rubiaceae 3): 811. 1991.

Tree or shrub. Montane forest, woodland/forest transition. 1900–2360 m.

Distribution: MR, NL, SL—*Gereau 4336*

Rytigynia stolzii Robyns; Verdcourt, F.T.E.A. (Rubiaceae 3): 812. 1991.

Shrub or tree. Riverine forest, montane forest. 1160–1980 m.

Distribution: MP, RM, SL—*Gereau 3685*

Rytigynia uhligii (K. Schum. & K. Krause) Verdc.; Verdcourt, F.T.E.A. (Rubiaceae 3): 818. 1991.

Tree or shrub. Disturbed montane forest, miombo woodland. 1550–2260 m.

Distribution: MP, NL, RM—*Gereau 7004*

Spermacoce dibrachiata Oliv.; Verdcourt, F.T.E.A. (Rubiaceae 1): 370. 1976.

Terrestrial herb. Montane grassland, secondary shrubland, woodland, abandoned cultivation. 1300–2140 m.

Distribution: MP, NL, SL—*Suleiman 122*

Spermacoce natalensis Hochst.; Verdcourt, F.T.E.A. (Rubiaceae 1): 359. 1976.

Terrestrial herb. Montane grassland and shrubland, scree slopes. 1760–2140 m.

Distribution: NL, SL—*Gereau 4238*

Spermacoce princeae (K. Schum.) Verdc.; Verdcourt, F.T.E.A. (Rubiaceae 1): 362. 1976.

Terrestrial herb. Montane forest and grassland. 1790–2140 m.

Distribution: NL, SL—*Gereau 4173*

Spermacoce pusilla Wall.; Verdcourt, F.T.E.A. (Rubiaceae 1): 356. 1976.

Terrestrial herb. Woodland. 1250–1800 m.

Distribution: MP—*P. Lovett 242*

Spermacoce sphaerostigma (A. Rich.) Oliv.; Verdcourt, F.T.E.A. (Rubiaceae 1): 367. 1976.

Terrestrial herb. Woodland, riverine forest. 1200–1600 m.

Distribution: MP, SL—*Gereau 4282*

Tarenna pavettoides (Harv.) Sim subsp. *affinis* (K. Schum.) Bridson; Bridson, F.T.E.A. (Rubiaceae 2): 589. 1988.

Shrub or tree. Montane forest. 1780 m.

Distribution: RM—*Kibure 1422*

Temnocalyx nodulosus Robyns; Verdcourt, F.T.E.A. (Rubiaceae 3): 761. 1991.

Shrub or tree. Bamboo thicket, montane forest edges. 2590–2850 m.

Distribution: MR, NL, RM—*Stolz 2296* (type: BM, EA, K, MO), *Abeid 2996*

Tricalysia acocantheroides K. Schum.; Robbrecht, F.T.E.A. (Rubiaceae 2): 556. 1988.

Shrub. Montane forest. 1900–2110 m.

Distribution: MP, MR, NL—*Gereau 4553*

Tricalysia longituba De Wild. var. *longituba*; Robbrecht, 1982: 320.

Shrub. Riverine forest. 980 m.

Distribution: SL—*Gereau 5145*

Note: New to East Africa, from Zambia and Democratic Republic of Congo.

Tricalysia ngalaensis Robbr.; Robbrecht, 1979: 277.

Shrub. Riverine forest. 580 m.

Distribution: SL—*Gereau 3815*

Note: New to East Africa, from Malawi.

Tricalysia pallens Hiern; Robbrecht, F.T.E.A. (Rubiaceae 2): 548. 1988.

Tree. Lower montane forest. 1470 m.

Distribution: NL—*Bracebridge 110*

Tricalysia verdcourtiana Robbr.; Robbrecht, F.T.E.A. (Rubiaceae 2): 549. 1988.

Shrub or treelet. Montane forest. 1780–1900 m.

Distribution: NL, RM—*Abeid 2772*

Vangueria infausta Burch.; Verdcourt, F.T.E.A. (Rubiaceae 3): 851. 1991.

Tree or shrub. Woodland, riverine forest, abandoned cultivation. 830–2000 m.

Distribution: MP, SL—*Gereau 4785*

Vangueria madagascariensis J. F. Gmel.; Verdcourt, F.T.E.A. (Rubiaceae 3): 849. 1991.

Shrub or tree. Montane forest, miombo woodland. 1450–2110 m.

Distribution: MP, MR—*Gereau 4570*

Vangueria volkensii K. Schum.; Verdcourt, F.T.E.A. (Rubiaceae 3): 854. 1991.

Tree. Riverine forest. 2000 m.

Distribution: MP—*J. Lovett 4985*

RUTACEAE

4 genera and 7 species

Clausena anisata (Willd.) Hook. f. ex Benth.; Kokwaro, F.T.E.A. (Rutaceae): 49. 1982.

Shrub or small tree. Montane forest, riverine forest, forest clearings, woodland. 1400–2500 m.

Distribution: KP, MP, RM, SL—*Abeid 2883*

Toddalia asiatica (L.) Lam.; Kokwaro, F.T.E.A. (Rutaceae): 3. 1982.

Liana or lianescent shrub. Tree clumps in grassland, montane forest. 1700–2360 m.

Distribution: MP, MR, RM—*Abeid 2927*

Vepris hanangensis (Kokwaro) Mziray; Lovett et al., 2006: 236.

Shrub. Riverine forest. 1300–1400 m.

Distribution: MP—*J. Lovett 5009*

Vepris nobilis (Delile) Mziray; Lovett et al., 2006: 236.

Tree. Riverine forest. 1320 m.

Distribution: MP—*Gereau 6118*

Vepris simplicifolia (Engl.) Mziray; Lovett et al., 2006: 237.

Tree. Montane forest. 1920–2150 m.

Distribution: NL—*Bracebridge 58*

Vepris stolzii I. Verd.; Lovett et al., 2006: 237.

Tree. Montane forest. 1750–1900 m.

Distribution: NL, RM—*Gereau 4187*

Zanthoxylum deremense (Engl.) Kokwaro; Kokwaro, F.T.E.A. (Rutaceae): 44. 1982.

Shrub or tree. Montane forest. 1720–2030 m.

Distribution: NL, RM—*Abeid 2882*

SANTALACEAE

3 genera and 7 species

Osyridicarpos schimperianus (Hochst. ex A. Rich.) A. DC.; Polhill, F.T.E.A. (Santalaceae): 21. 2005.

Scrambling shrub or liana. Riverine forest, woodland. 1150–1250 m.

Distribution: MP, SL—*Kayombo 2456*

Osyris lanceolata Hochst & Steud.; Polhill, F.T.E.A. (Santalaceae): 23. 2005.

Shrub or tree. Riverine forest, woodland, montane thickets. 980–2140 m.

Distribution: KP, MP, SL—*Gereau 7005*

Thesium fimbriatum A. W. Hill; Polhill, F.T.E.A. (Santalaceae): 14. 2005.

Terrestrial herb. Miombo woodland, montane grassland. 1400–1940 m.

Distribution: MP, SL—*Gereau 4664*

Thesium kilimandscharicum Engl.; Polhill, F.T.E.A. (Santalaceae): 7. 2005.

Suffrutex. Montane grassland, rocky ridges. 2500–2960 m.

Distribution: KP, RM—*Taylor 8522*

Thesium subaphyllum Engl.; Polhill, F.T.E.A. (Santalaceae): 12. 2005.

Terrestrial herb. Miombo woodland. 1790 m.

Distribution: SL—*Gereau 3899*

Thesium triflorum Thunb. ex L. f.; Polhill, F.T.E.A. (Santalaceae): 6. 2005.

Straggling shrub. Montane forest. 2040 m.

Distribution: SL—*Gereau 4739*

Thesium unyikense Engl.; Polhill, F.T.E.A. (Santalaceae): 16. 2005.

Terrestrial herb. Woodland. 1400–1500 m.

Distribution: MP—*J. Lovett 3401*

SAPINDACEAE

5 genera and 12 species

Allophylus abyssinicus (Hochst.) Radlk.; Davies & Verdcourt, F.T.E.A. (Sapindaceae): 78. 1998.

Tree. Montane forest. 1750–2110 m.

Distribution: MR, RM—*Abeid 2909*

Allophylus africanus P. Beauv. var. *griseotomentosus* (Gilg) Verdc.; Davies & Verdcourt, F.T.E.A. (Sapindaceae): 82. 1998.

Tree or shrub. Riverine forest. 580–1500 m.

Distribution: SL—*Gereau 4393*

Allophylus chaunostachys Gilg; Davies & Verdcourt, F.T.E.A. (Sapindaceae): 84. 1998.

Tree or shrub, sometimes lianescent. Montane forest, woodland. 1530–2450 m.

Distribution: MR, NL, RM, SL—*Suleiman 107*

Allophylus congolanus Gilg; Davies & Verdcourt, F.T.E.A. (Sapindaceae): 92. 1998.

Shrub, sometimes lianescent. Riverine forest, miombo woodland. 1100–1530 m.

Distribution: MP, SL—*Mwangoka 660*

Allophylus fulvotomentosus Gilg; Davies & Verdcourt, F.T.E.A. (Sapindaceae): 82. 1998.

Tree or shrub, sometimes lianescent. Woodland. 1100–1260 m.

Distribution: MP—*Kayombo 2358*

Allophylus richardsiae Exell; Davies & Verdcourt, F.T.E.A. (Sapindaceae): 92. 1998.

Tree or shrub. Woodland, riverine forest, abandoned cultivation. 1250–1600 m.

Distribution: MP—*J. Lovett 3393*

Allophylus tanzaniensis F. G. Davies; Davies & Verdcourt, F.T.E.A. (Sapindaceae): 87. 1998.

Shrub. Riverine forest. 1200 m.

Distribution: MP—*J. Lovett 4307*

Blighia unijugata Baker; Davies & Verdcourt, F.T.E.A. (Sapindaceae): 25. 1998.

Tree. Lakeshore bushland, riverine forest. 474–840 m.

Distribution: SL—*Mwasumbi 16587*

Deinbollia borbonica Scheff. f. *glabrata* Radlk.; Davies & Verdcourt, F.T.E.A. (Sapindaceae): 71. 1998.

Shrub. Riverine forest. 880 m.

Distribution: SL—*Kibure 1452*

Deinbollia kilimandscharica Taub. var. *adusta* (Radlk.) Verdc.; Davies & Verdcourt, F.T.E.A. (Sapindaceae): 71. 1998.

Tree. Montane forest. 1730 m.

Distribution: NL—*Bracebridge 55*

Dodonaea viscosa Jacq. var. *angustifolia* (L. f.) Benth.; Davies & Verdcourt, F.T.E.A. (Sapindaceae): 9. 1998.

Shrub or tree. Montane thickets, secondary forest, riverine forest. 1500–2510 m.

Distribution: NL, RM, SL—*Suleiman 139*

Paullinia pinnata L.; Davies & Verdcourt, F.T.E.A. (Sapindaceae): 102. 1998.

Liana. Riverine forest, lakeshore bushland, abandoned cultivation. 474–1400 m.

Distribution: MP, SL—*Kayombo 1024*

SAPOTACEAE

5 genera and 9 species

Chrysophyllum gorungosanum Engl.; Hemsley, F.T.E.A. (Sapotaceae): 10. 1968.

Tree. Montane forest. 1840 m.

Distribution: RM—*Kibure 1389*

Englerophytum magalismontanum (Sond.) T. D. Penn.; Lovett et al., 2006: 254.

Shrub or small tree. Miombo woodland, montane forest. 980–1800 m.

Distribution: NL, SL—*Harder 1528*

Englerophytum natalense (Sond.) T. D. Penn.; Lovett et al., 2006: 255.

Tree. Riverine forest. 880–1400 m.

Distribution: MP, SL—*Kibure 1453*

Mimusops kummel Bruce ex A. DC.; Hemsley, F.T.E.A. (Sapotaceae): 54. 1968.

Tree. Riverine forest. 474–480 m.

Distribution: RM, SL—*Mwasumbi 16596*

Mimusops obtusifolia Lam.; Lovett et al., 2006: 259.

Shrub. Riverine forest, miombo woodland. 620–980 m.

Distribution: SL—*Gereau 5151*

Mimusops zeyheri Sond.; Hemsley, F.T.E.A. (Sapotaceae): 56. 1968.

Tree. Riverine forest. 1350 m.

Distribution: MP—*J. Lovett 4667*

Pouteria adolfi-friedericii (Engl.) A. Meeuse subsp. *australis* (J. H. Hemsl.) L. Gaut.; Lovett et al., 2006: 261.

Tree. Riverine forest, montane forest. 1840–2200 m.

Distribution: MP, RM—*Semsei 1532* (type: EA, K), *Kayombo 1078*

Pouteria alnifolia (Baker) Roberty var. *alnifolia*; Lovett et al., 2006: 263.

Tree. Disturbed montane forest.

Distribution: RM—*Ndangalasi 960*

Synsepalum brevipes (Baker) T. D. Penn.; Lovett et al., 2006: 265.

Tree. Riverine forest. 830–920 m.

Distribution: SL—*Kibure 1476*

SCROPHULARIACEAE

13 genera and 22 species

Alectra sessiliflora (Vahl) Kuntze; Ghazanfar, F.T.E.A. (Scrophulariaceae): 112. 2008.

Terrestrial herb. Roadsides in cultivation, riverine forest, woodland. 1250–2200 m.

Distribution: KP, MP, MR—*Gereau 4516*

Buchnera cryptocephala (Baker) Philcox; Philcox, F.T.E.A. (Scrophulariaceae): 126. 2008.

Terrestrial herb. Woodland, montane grassland. 1400–2300 m.

Distribution: MP, SL—*J. Lovett 4159*

Buchnera decandollei Govaerts; Govaerts, 1996: 14.

Terrestrial herb. Wet grassland. 1250–1400 m.

Distribution: MP—*Kayombo 971*

Buchnera lastii Engl. subsp. *lastii*; Philcox, F.T.E.A. (Scrophulariaceae): 133. 2008.

Terrestrial herb. 2250 m.

Distribution: MP—*J. Lovett 4841*

Buchnera quadrifaria Baker; Philcox, F.T.E.A. (Scrophulariaceae): 129. 2008.

Terrestrial herb. Montane grassland. 1940 m.

Distribution: KP, SL—*Gereau 4779*

Buchnera rungwensis Engl.; Philcox, F.T.E.A. (Scrophulariaceae): 129. 2008.

Terrestrial herb. Montane grassland, rocky ridges. 1940–2960 m.

Distribution: KP, RM, SL—*Stolz 2439* (type: K), *Suleiman 36*

Buchnera speciosa Skan; Philcox, F.T.E.A. (Scrophulariaceae): 136. 2008.

Terrestrial herb. Riverine forest. 1500 m.

Distribution: MP, SL—*Gereau 4800*

Buchnera trilobata Skan vel sp. aff.; Philcox, Fl.

Zamb. 8(2): 102. 1990.

Terrestrial herb. Woodland. 1400–1500 m.

Distribution: MP—*J. Lovett 3408*

Cycnium adonense E. Mey. ex Benth.; Hepper, F.T.E.A. (Scrophulariaceae): 168. 2008.

Terrestrial herb. Woodland, scree slopes. 1200–2000 m.

Distribution: MP, SL—*Gereau 3638*

Diclis ovata Benth.; Hepper & Ghazanfar, F.T.E.A. (Scrophulariaceae): 20. 2008.

Terrestrial herb. Abandoned cultivation, open mountain slopes, riverine forest. 1250–1980 m.

Distribution: MP, SL—*Gereau 3642*

Halleria lucida L.; Hepper, F.T.E.A. (Scrophulariaceae): 26. 2008.

Shrub or tree. Riverine forest, montane forest. 1780–2500 m.

Distribution: MP, MR, NL, RM, SL—*Harder 1477*

Hebenstretia angolensis Rolfe; Ghazanfar, F.T.E.A. (Scrophulariaceae): 197. 2008.

Terrestrial herb or shrub. Woodland, grassland. 1300–2600 m.

Distribution: KP, MP—*Suleiman 41*

Hedbergia abyssinica (Hochst. ex Benth.) Molau; Philcox, F.T.E.A. (Scrophulariaceae): 195. 2008.

Terrestrial herb. Montane forest, montane grassland. 2200–2700 m.

Distribution: KP, MR, NL—*Gereau 4604*

Jamesbrittenia micrantha (Klotzsch) Hilliard; Ghazanfar, F.T.E.A. (Scrophulariaceae): 32. 2008.

Terrestrial herb. Abandoned cultivation, riverine forest. 1250–1350 m.

Distribution: MP—*J. Lovett 3779*

Lindernia nummulariifolia (D. Don) Wettst.; Philcox, F.T.E.A. (Scrophulariaceae): 80. 2008.

Terrestrial herb. Abandoned cultivation. 1250–1600 m.

Distribution: MP—*J. Lovett 4023*

Lindernia rupestris Engl.; Philcox, F.T.E.A. (Scrophulariaceae): 72. 2008.

Terrestrial herb. Miombo woodland. 1200–1800 m.

Distribution: MP—*P. Lovett 247*

Micrargeria filiformis (Schumach. & Thonn.) Hutch. & Dalziel; Ghazanfar, F.T.E.A. (Scrophulariaceae): 187. 2008.

Terrestrial herb. Wet grassland. 1250 m.

Distribution: MP—*J. Lovett 4570*

Sopubia mannii Skan var. *linearifolia* O. J. Hansen; Philcox, F.T.E.A. (Scrophulariaceae): 178. 2008.

Terrestrial herb. Montane grassland. 1940–2840 m.

Distribution: KP, MP, SL—*Richards 14183* (type: EA, K), *Suleiman 168*

Sopubia ramosa (Hochst.) Hochst.; Philcox, F.T.E.A. (Scrophulariaceae): 180. 2008.

Terrestrial herb or suffrutex. Riverine forest, woodland. 1300–2000 m.

Distribution: MP, SL—*Gereau 4801*

Striga asiatica (L.) Kuntze; Hepper, F.T.E.A. (Scrophulariaceae): 148. 2008.

Terrestrial herb. Woodland. 1250 m.

Distribution: MP—*J. Lovett 4331*

Veronica abyssinica Fresen.; Hepper, F.T.E.A. (Scrophulariaceae): 102. 2008.

Terrestrial herb. Montane grassland, montane forest edges. 1900–2640 m.

Distribution: KP, MR, NL, SL—*Suleiman 116*

Veronica glandulosa Hochst. ex Benth.; Hepper, F.T.E.A. (Scrophulariaceae): 104. 2008.

Terrestrial herb. Montane grassland. 2570 m.

Distribution: KP—*Gereau 3092*

SIMAROUBACEAE

1 genus and 1 species

Harrisonia abyssinica Oliv.; Stannard, F.T.E.A. (Simaroubaceae): 2. 2000.

Shrub. Lakeshore shrubland. 474–490 m.

Distribution: SL—*Mwasumbi 16553*

SMILACACEAE

1 genus and 2 species

Smilax anceps Willd.; Cowley, F.T.E.A. (Smilacaceae): 2. 1989.

Liana. Montane forest and thickets, riverine forest, edges of cultivation. 1250–2140 m.

Distribution: MP, NL, SL—*Gereau 4202*

Smilax aspera L.; Cowley, F.T.E.A. (Smilacaceae): 2. 1989.

Liana. Montane thickets, forest edges. 1840–2120 m.

Distribution: SL—*Gereau 4315*

SOLANACEAE

5 genera and 11 species

Cestrum aurantiacum Lindl.; Gonçalves, Fl. Zamb. 8(4): 11. 2005.

Shrub. Riverine forest, abandoned cultivation. 1250–1400 m.

Distribution: MP—*Kayombo 1028*

Discopodium penninervium Hochst.; Gonçalves, Fl. Zamb. 8(4): 63. 2005.

Shrub. Montane forest. 1720–2350 m

Distribution: MR, RM—*Mwasumbi 16225*

Nicandra physalodes (L.) Gaertn.; Gonçalves, Fl. Zamb. 8(4): 26. 2005.

Terrestrial herb. Weed of cultivation. 1250 m.

Distribution: MP—*Kayombo 2466*

Physalis peruviana L.; Gonçalves, Fl. Zamb. 8(4): 47. 2005.

Terrestrial herb. Riverine forest. 1500 m.

Distribution: MP—*J. Lovett 4694*

Solanum aculeastrum Dunal; Gonçalves, Fl. Zamb. 8(4): 107. 2005.

Shrub. Montane forest. 1870 m.

Distribution: RM—*WCS SHCP 117*

Solanum aculeatissimum Jacq.; Gonçalves, Fl. Zamb. 8(4): 118. 2005.

Shrub or perennial herb. Rocky ridges, riverine forest. 1850–2600 m.

Distribution: MP, SL—*Gereau 3494*

Solanum americanum Mill.; Gonçalves & Edmonds, Fl. Zamb. 8(4): 81. 2005.

Terrestrial herb. Riverine forest, montane forest, abandoned cultivation. 1250–2230 m.

Distribution: MP, SL—*Gereau 4426*

Solanum anguivi Lam.; Gonçalves, Fl. Zamb. 8(4): 93. 2005.

Shrub. Montane forest, roadsides in cultivation, grassland, riverine forest. 1210–2300 m.

Distribution: MP, MR, NL, RM, SL—*Gereau 4515*

Solanum incanum L.; Gonçalves, Fl. Zamb. 8(4): 112. 2005.

Shrub or suffrutex. Roadsides, edges of cultivation, woodland, heathland. 474–2700 m.

Distribution: KP, MP, SL—*Gereau 3786*

Solanum schumannianum Dammer; Gonçalves, Fl. Zamb. 8(4): 90. 2005.

Shrub. Montane forest. 1900–2370 m.

Distribution: MR, NL, RM, SL—*Mwasumbi 16356*

Solanum terminale Forssk.; Gonçalves, Fl. Zamb. 8(4): 77. 2005.

Liana or shrub. Montane forest, forest edges. 2260–2500 m.

Distribution: MP, SL—*Gereau 3522*

STERCULIACEAE

5 genera and 7 species

Cola greenwayi Brenan vel sp. aff.; Cheek, F.T.E.A. (Sterculiaceae): 42. 2007.

Tree. Montane forest, riverine forest. 890–1630 m.

Distribution: RM, SL—*Bracebridge 41*

Dombeya acutangula Cav.; Cheek, F.T.E.A. (Sterculiaceae): 62. 2007.

Shrub or tree. Riverine forest, roadsides. 1150–1350 m.

Distribution: MP, NL—*J. Lovett 4660*

Dombeya burgessiae Gerrard ex Harv.; Cheek, F.T.E.A. (Sterculiaceae): 64. 2007.

Shrub or tree. Woodland, riverine forest, montane forest, secondary forest. 1250–2300 m.

Distribution: MP, MR, RM—*Abeid 2911*

Dombeya torrida (J. F. Gmel.) Bamps subsp. *erythroleuca* (K. Schum.) Seyani; Cheek, F.T.E.A. (Sterculiaceae): 67. 2007.

Shrub or tree. Montane forest, montane grassland. 1900–2350 m.

Distribution: MR, RM, SL—*Gereau 4765*

Leptonychia usambarensis K. Schum.; Dorr, F.T.E.A. (Sterculiaceae): 102. 2007.

Tree. Montane forest. 1700 m.

Distribution: NL—*Bracebridge 64*

Sterculia quinqueloba (Garcke) K. Schum.; Cheek, F.T.E.A. (Sterculiaceae): 12. 2007.

Tree. Woodland. 1250–1300 m.

Distribution: MP—*Kayombo 1096*

Waltheria indica L.; Dorr, F.T.E.A. (Sterculiaceae): 127. 2007.

Herb or suffrutex. Miombo woodland. 1200–1250 m.

Distribution: MP—*J. Lovett 4067*

TACCACEAE

1 genus and 1 species

Tacca leontopetaloides (L.) Kuntze; Carter, F.T.E.A. (Taccaceae): 1. 1962.

Terrestrial herb. Woodland, lakeshore forest. 474–1250 m.

Distribution: MP, SL—*Gereau 3801*

THEACEAE

1 genus and 1 species

Ficalhoa laurifolia Hiern; Verdcourt, F.T.E.A. (Theaceae): 8. 1962.

Tree. Montane forest, riverine forest. 1640–2600 m.

Distribution: MR, NL, RM, SL—*Gereau 4609*

THYMELAEACEAE

4 genera and 9 species

Dais cotinifolia L.; Peterson, F.T.E.A. (Thymelaeaceae): 16. 1978.

Tree. Woodland. 2080 m.

Distribution: SL—*Gereau 4016*

Gnidia chrysantha (Solms) Gilg; Petersen, F.T.E.A. (Thymelaeaceae): 26. 1978.

Terrestrial herb or suffrutex. Miombo woodland, grassland. 1450–1700 m.

Distribution: MP, RM—*Gereau 6980*

Gnidia fastigiata Rendle; Petersen, F.T.E.A. (Thymelaeaceae): 21. 1978.

Terrestrial herb or suffrutex. Montane grassland, rocky ridges. 2000–2800 m.

Distribution: KP, MP, RM, SL—*Gereau 5137*

Gnidia glauca (Fresen.) Gilg; Peterson, F.T.E.A. (Thymelaeaceae): 32. 1978.

Shrub. Montane forest edges, wooded grassland. 2100–2685 m.

Distribution: KP, MP, NL—*Gereau 4641*

Gnidia involucrata Steud. ex A. Rich.; Peterson, F.T.E.A. (Thymelaeaceae): 25. 1978.

Terrestrial herb or suffrutex. Woodland. 1400–1800 m.

Distribution: MP—*J. Lovett 3524*

Gnidia kraussiana Meisn.; Peterson, F.T.E.A. (Thymelaeaceae): 29. 1978.

Terrestrial herb. Montane grassland. 2230 m.

Distribution: SL—*Gereau 3602*

Gnidia usafuae Gilg; Peterson, F.T.E.A. (Thymelaeaceae): 26. 1978.

Shrub. Montane grassland and forest edges.

Distribution: MP, RM—*Semsei 1541*

Peddiea polyantha Gilg; Peterson, F.T.E.A. (Thymelaeaceae): 11. 1978.

Shrub or tree. Montane forest. 1900–2260 m.

Distribution: MR, NL, RM, SL—*Abeid 2846*

Synaptolepis alternifolia Oliv.; Peterson, F.T.E.A. (Thymelaeaceae): 6. 1978.

Liana. 1740 m.

Distribution: RM—*WCS SHCP 28*

TILIACEAE

4 genera and 16 species

Corchorus tridens L.; Andrews, F.T.E.A. (Tiliaceae): 113. 2001.

Terrestrial herb. Riverine forest. 1350 m.

Distribution: MP—*J. Lovett 4657*

Grewia bicolor Juss.; Whitehouse, F.T.E.A. (Tiliaceae): 42. 2001.

Tree. Riverine forest. 580 m.

Distribution: SL—*Gereau 3822*

Grewia flavescens Juss.; Whitehouse, F.T.E.A. (Tiliaceae): 33. 2001.

Shrub or liana. Woodland, grassland, riverine forest, abandoned cultivation. 1200–1600 m.

Distribution: MP—*Kayombo 2413*

Grewia similis K. Schum.; Whitehouse, F.T.E.A. (Tiliaceae): 13. 2001.

Shrub or small tree. Riverine forest. 880 m.

Distribution: SL—*Kibure 1454*

Grewia stolzii Ulbr.; Whitehouse, F.T.E.A. (Tiliaceae): 22. 2001.

Treelet. Opening in montane forest. 1920 m.

Distribution: MR, SL—*Gereau 6126*

Sparrmannia ricinocarpa (Eckl. & Zeyh.) Kuntze; Verdcourt, F.T.E.A. (Tiliaceae): 93. 2001.

Straggling terrestrial herb or shrub. Road banks, montane grassland, montane forest. 2040–2700 m.

Distribution: KP, MP, MR, NL, SL—*Suleiman 66*

Triumfetta amuletum Sprague; Cheek, F.T.E.A. (Tiliaceae): 70. 2001.

Trailing shrub. Grassland, miombo woodland. 1200–1400 m.

Distribution: MP—*Kayombo 1015*

Triumfetta angolensis Sprague & Hutch.; Wild, Fl. Zamb. 2: 71. 1963.

Terrestrial herb. Miombo woodland. 1800 m.

Distribution: MP—*P. Lovett 20*

Note: New to East Africa, from Zambia, Malawi, Mozambique, Zimbabwe, South Africa, and Angola.

Triumfetta annua L.; Cheek, F.T.E.A. (Tiliaceae): 79. 2001.

Terrestrial herb. Woodland. 1250–2080 m.

Distribution: MP, MR—*Mwasumbi 16191*

Triumfetta brachyceras K. Schum.; Cheek, F.T.E.A. (Tiliaceae): 89. 2001.

Shrub. Montane forest edges. 1770 m.

Distribution: RM—*WCS SHCP 48*

Triumfetta pentandra A. Rich.; Cheek, F.T.E.A. (Tiliaceae): 85. 2001.

Terrestrial herb. Riverine forest. 1500 m.

Distribution: SL—*Gereau 4791*

Triumfetta pilosa Roth; Cheek, F.T.E.A. (Tiliaceae): 91. 2001.

Shrub, subshrub, herb. Miombo woodland, riverine forest, disturbed montane forest, montane grassland. 1530–1940 m.

Distribution: MP, RM, SL—*Gereau 4386*

Triumfetta rhomboidea Jacq.; Cheek, F.T.E.A. (Tiliaceae): 83. 2001.

Terrestrial herb, suffrutex, shrub. Montane forest, riverine forest, woodland, grassland, edges of cultivation. 1200–2200 m.

Distribution: MP, MR, SL—*Suleiman 90*

Triumfetta setulosa Mast.; Cheek, F.T.E.A. (Tiliaceae): 76. 2001.

Terrestrial herb. Riverine forest. 1210 m.

Distribution: MR, SL—*Gereau 4255*

Triumfetta tomentosa Bojer; Cheek, F.T.E.A. (Tiliaceae): 90. 2001.

Shrub. Woodland, riverine forest. 1200–1600 m.

Distribution: MP—*J. Lovett 4419*

Triumfetta welwitschii Mast. var. *welwitschii*; Cheek, F.T.E.A. (Tiliaceae): 74. 2001.

Terrestrial herb. Montane grassland, woodland. 1250–2250 m.

Distribution: MP—*J. Lovett 4870*

TURNERACEAE

1 genus and 2 species

Tricliceras lobatum (Urb.) R. Fern.; Fernandes, Fl. Zamb. 4: 363. 1978.

Terrestrial herb. Roadsides. 474–490 m.

Distribution: SL—*Gereau 3797*

Tricliceras longipedunculatum (Mast.) R. Fern. var. *longepidunculatum*; Fernandes, Fl. Zamb. 4: 359. 1978.

Terrestrial herb. Woodland, riverine forest. 1200–1400 m.

Distribution: MP—*J. Lovett 5021*

ULMACEAE

2 genera and 3 species

Celtis africana Burm. f.; Polhill, F.T.E.A. (Ulmaceae): 4. 1966.

Tree. Montane forest. 1800 m.

Distribution: NL, RM—*Bracebridge 54*

Celtis gomphophylla Baker; Lovett et al., 2006: 279.

Tree. Riverine forest. 1320–1480 m.

Distribution: MP—*Lovett 4894*

Trema orientalis (L.) Blume; Polhill, F.T.E.A. (Ulmaceae): 10. 1966.

Tree. Miombo woodland, riverine forest, montane forest, abandoned cultivation. 620–1830 m.

Distribution: MP, NL, RM, SL—*Harder 1522*

URTICACEAE

9 genera and 13 species

Boehmeria macrophylla Hornem.; Friis, F.T.E.A. (Urticaceae): 44. 1989.

Shrub. Riverine forest, montane forest. 1210–2040 m.

Distribution: MP, NL, RM, SL—*Gereau 4277*

Droguetia iners (Forssk.) Schweinf. subsp. *iners*; Friis, F.T.E.A. (Urticaceae): 57. 1989.

Straggling terrestrial herb. Tree clumps in montane grassland, hedge beside cultivated field. 2080–2360 m.

Distribution: MR, NL—*Gereau 3749*

Elatostema monticola Hook. f.; Friis, F.T.E.A. (Urticaceae): 41. 1989.

Terrestrial herb. Montane forest. 2460 m.

Distribution: NL, RM—*Gereau 4112*

Elatostema welwitschii Engl.; Friis, F.T.E.A. (Urticaceae): 39. 1989.

Terrestrial herb. Riverine forest. 1790 m.

Distribution: NL—*Gereau 4160*

Girardinia diversifolia (Link) Friis; Friis, F.T.E.A. (Urticaceae): 13. 1989.

Terrestrial herb. Woodland, riverine forest. 1275–1400 m.

Distribution: MP—*Kayombo 1004*

Laportea alatipes Hook. f.; Friis, F.T.E.A. (Urticaceae): 16. 1989.

Terrestrial herb. Montane forest. 2400–2460 m.

Distribution: MP, NL—*Gereau 4122*

Laportea interrupta (L.) Chew; Friis, F.T.E.A. (Urticaceae): 17. 1989.

Terrestrial herb. Abandoned cultivation. 1250 m.

Distribution: MP—*J. Lovett 4037*

Pilea rivularis Wedd.; Friis, F.T.E.A. (Urticaceae): 29. 1989.

Terrestrial herb. Montane forest. 2120–2400 m.

Distribution: MP, SL—*Gereau 4331*

Pilea tetraphylla (Steud.) Blume; Friis, F.T.E.A. (Urticaceae): 27. 1989.

Terrestrial herb. Montane forest, riverine forest. 1790–1980 m.

Distribution: MR, NL—*Gereau 4153*

Pilea usambarensis Engl.; Friis, F.T.E.A. (Urticaceae): 36. 1989.

Terrestrial herb. Montane forest, riverine forest. 1790–1850 m.

Distribution: NL, SL—*Gereau 4164*

Pouzolzia mixta Solms; Friis, F.T.E.A. (Urticaceae): 48. 1989.

Shrub. Lakeshore forest and shrubland. 480–495 m.

Distribution: RM, SL—*Gereau 3647*

Procris crenata C. B. Rob.; Friis, F.T.E.A. (Urticaceae): 42. 1989.

Epiphyte. Riverine forest. 1710–1750 m.

Distribution: RM—*Mwasumbi 16334*

Urera hypselodendron (Hochst. ex A. Rich.) Wedd.; Friis, F.T.E.A. (Urticaceae): 9. 1989.

Liana. Montane forest and grassland. 1780–2300 m.

Distribution: KP, MR, RM—*Mwasumbi 16307*

VALERIANACEAE

1 genus and 1 species

Valeriana capensis Thunb.; Kokwaro, F.T.E.A. (Valerianaceae): 6. 1968.

Terrestrial herb. Montane grassland. 2360–2550 m.

Distribution: KP, MR—*Gereau 3775*

VERBENACEAE

6 genera and 17 species

Clerodendrum eriophyllum Gürke; Verdcourt, F.T.E.A. (Verbenaceae): 120. 1992.

Shrub. Woodland. 1200–1250 m.

Distribution: MP—*J. Lovett 4232*

Clerodendrum frutectorum S. Moore; Verdcourt, F.T.E.A. (Verbenaceae): 103. 1992.

Shrub. Riverine forest, woodland. 1200–1350 m.

Distribution: MP—*J. Lovett 4408*

Clerodendrum johnstonii Oliv. subsp. *johnstonii*; Verdcourt, F.T.E.A. (Verbenaceae): 118. 1992.

Scrambling shrub. Tree clumps in grassland, montane forest. 1960–2360 m.

Distribution: MR, SL—*Gereau 3751*

Lantana camara L.; Verdcourt, F.T.E.A. (Verbenaceae): 39. 1992.

Shrub. Montane forest. 2010 m.

Distribution: RM—*WCS SHCP 182*

Lantana trifolia L.; Verdcourt, F.T.E.A. (Verbenaceae): 46. 1992.

Shrub. Miombo woodland. 1800 m.

Distribution: MP—*P. Lovett 42*

Lantana ukambensis (Vatke) Verdc.; Verdcourt, F.T.E.A. (Verbenaceae): 43. 1992.

Terrestrial herb or suffrutex. Woodland. 1200 m.

Distribution: MP—*J. Lovett 4255*

Lippia abyssinica (Otto & A. Dietr.) Cufod.; Verdcourt, F.T.E.A. (Verbenaceae): 29. 1992.

Shrub or perennial herb. Riverine forest edges, montane grassland. 1500–2140 m.

Distribution: SL—*Gereau 4401*

Lippia javanica (Burm. f.) Spreng.; Verdcourt, F.T.E.A. (Verbenaceae): 30. 1992.

Shrub. Roadsides in cultivation, montane forest. 1300–2200 m.

Distribution: MP, MR, RM—*Gereau 4517*

Lippia plicata Baker; Verdcourt, F.T.E.A. (Verbenaceae): 34. 1992.

Shrub. Woodland, riverine forest, montane grassland, montane forest, abandoned cultivation. 1200–2010 m.

Distribution: MP, RM, SL—*Gereau 4655*

Premna senensis Klotzsch; Verdcourt, F.T.E.A. (Verbenaceae): 78. 1992.

Shrub, sometimes lianescent. Woodland, lakeshore forest, rocky beaches. 474–1200 m.

Distribution: MP, SL—*Gereau 3654*

Rotheca alata (Gürke) Verdc.; Fernandes & Verdcourt, 2000: 148.

Shrub. Abandoned cultivation. 2000 m.

Distribution: MP—*P. Lovett 253*

Rotheca myricoides (Hochst.) Steane & Mabb. subsp. *austromonticola* (Verdc.) Verdc.; Fernandes & Verdcourt, 2000: 151.

Shrub or subshrubby terrestrial herb. Montane grassland, woodland, *Cupressus* plantations. 1250–2550 m.

Distribution: MP, MR, RM—*Richards 14342* (type: K), *Abeid 2834*

Rotheca myricoides (Hochst.) Steane & Mabb. subsp. *myricoides*; Fernandes & Verdcourt, 2000: 151.

Shrub or treelet. Montane forest, woodland, lakeshore forest, grassland, abandoned cultivation. 474–2350 m.

Distribution: MP, MR, RM, SL—*Suleiman 150*

Vitex buchananii Baker ex Gürke; Verdcourt, F.T.E.A. (Verbenaceae): 54. 1992.

Shrub. Montane forest. 1760 m.

Distribution: RM—*WCS SHCP 338*

Vitex doniana Sweet; Verdcourt, F.T.E.A. (Verbenaceae): 62. 1992.

Tree. Riverine forest, woodland, abandoned cultivation. 580–1600 m.

Distribution: MP, SL—*Kayombo 973*

Vitex fischeri Gürke; Verdcourt, F.T.E.A. (Verbenaceae): 59. 1992.

Tree. Abandoned cultivation. 1250 m.

Distribution: MP—*J. Lovett 3778*

Vitex mombassae Vatke; Verdcourt, F.T.E.A. (Verbenaceae): 64. 1992.

Tree. Woodland. 1250–1400 m.

Distribution: MP—*Kayombo 1095*

Vitex payos (Lour.) Merr. var. *glabrescens* (W. Piep.) Moldenke; Verdcourt, F.T.E.A. (Verbenaceae): 63. 1992.

Tree. Riverine forest. 580 m.

Distribution: SL—*Gereau 3810*

VIOLACEAE

2 genera and 2 species

Hybanthus enneaspermus (L.) F. Muell. var. *latifolius* (De Wild.) Engl.; Grey-Wilson, F.T.E.A. (Violaceae): 34. 1986.

Shrub or subshrub. Montane forest. 1790–2400 m.

Distribution: NL, RM—*Gereau 4151*

Viola abyssinica Steud. ex Oliv.; Grey-Wilson, F.T.E.A. (Violaceae): 25. 1986.

Terrestrial herb. Montane forest, montane

grassland. 2040–2750 m.

Distribution: KP, MR, NL, RM, SL—*Gereau 4723*

Note: Most collections are somewhat intermediate with *V. eminii* (Engl.) R. E. Fr.

VISCACEAE

1 genus and 4 species

Viscum congdonii Polhill & Wiens; Polhill & Wiens, F.T.E.A. (Viscaceae): 19. 1999.

Parasitic shrub. Montane forest and woodland. 2230–2350 m.

Distribution: KP, MR—*Wiens 6985* (type: BR, DSM, EA, K, MO, SRGH), *Mwasumbi 16173*

Viscum goetzei Engl.; Polhill & Wiens, F.T.E.A. (Viscaceae): 18. 1999.

Parasitic shrub. Montane forest. 2350 m.

Distribution: KP—*Goetze 1273* (type: B, BR, K), *Wiens 6986*

Viscum loranthicola Polhill & Wiens; Polhill & Wiens, F.T.E.A. (Viscaceae): 16. 1999.

Parasitic shrub. Woodland. 1310 m.

Distribution: MP—*Gereau 6101*

Viscum shirense Sprague; Polhill & Wiens, F.T.E.A. (Viscaceae): 23. 1999.

Parasitic shrub. Montane forest. 1920 m.

Distribution: SL—*Gereau 3619*

VITACEAE

5 genera and 25 species

Ampelocissus africana (Lour.) Merr.; Verdcourt, F.T.E.A. (Vitaceae): 5. 1993.

Herbaceous climber or liana. Riverine forest, lakeshore forest, woodland. 474–1400 m.

Distribution: MP, SL—*Gereau 4283*

Ampelocissus obtusata (Welw. ex Baker) Planch. subsp. *kirkiana* (Planch.) Wild & R. B. Drumm.; Verdcourt, F.T.E.A. (Vitaceae): 8. 1993.

Herbaceous climber. Woodland. 1200 m.

Distribution: MP—*J. Lovett 4051*

Ampelocissus obtusata (Welw. ex Baker) Planch. subsp. *obtusata*; Verdcourt, F.T.E.A. (Vitaceae): 7. 1993.

Terrestrial herb. Miombo woodland. 1200 m.

Distribution: MP—*J. Lovett 4044*

Cayratia gracilis (Guill. & Perr.) Suess.; Verdcourt, F.T.E.A. (Vitaceae): 138. 1993.

Herbaceous climber. Riverine forest, abandoned cultivation. 1250–1500 m.

Distribution: MP, SL—*Gereau 4395*

Cissus cornifolia (Baker) Planch.; Verdcourt, F.T.E.A. (Vitaceae): 36. 1993.

Shrub. *Acacia* woodland. 1250–1300 m.

Distribution: MP—*J. Lovett 3571*

Cissus faucicola Wild & R. B. Drumm.; Verdcourt, F.T.E.A. (Vitaceae): 47. 1993.

Herbaceous climber. Montane forest, riverine forest. 1250–2300 m.

Distribution: MP, SL—*Kayombo 2344*

Cissus integrifolia (Baker) Planch.; Verdcourt, F.T.E.A. (Vitaceae): 15. 1993.

Liana. Lakeshore shrubland and forest. 474–490 m.

Distribution: SL—*Gereau 3667*

Cissus oliveri (Engl.) Gilg; Verdcourt, F.T.E.A. (Vitaceae): 31. 1993.

Shrub. Montane forest. 1860 m.

Distribution: RM—*WCS SHCP 34*

Cissus rubiginosa (Welw. ex Baker) Planch.; Verdcourt, F.T.E.A. (Vitaceae): 26. 1993.

Herbaceous climber. Riverine forest. 580 m.

Distribution: SL—*Gereau 3811*

Cissus sciaphila Gilg vel sp. aff.; Verdcourt, F.T.E.A. (Vitaceae): 30. 1993.

Liana. Riverine forest. 1210 m.

Distribution: SL—*Gereau 4284*

Cyphostemma aff. *bambuseti* (Gilg & M. Brandt) Desc. ex Wild & R. B. Drumm.; Verdcourt, F.T.E.A. (Vitaceae): 119. 1993.

Herbaceous climber. Riverine forest. 1210–1600 m.

Distribution: MP, SL—*Gereau 4289*

Cyphostemma buchananii (Planch.) Desc. ex Wild & R. B. Drumm.; Verdcourt, F.T.E.A. (Vitaceae): 99. 1993.

Herbaceous climber or prostrate terrestrial herb. Riverine forest, montane forest. 1780–2140 m.

Distribution: RM, SL—*Suleiman 94*

Cyphostemma crotalarioides (Planch.) Desc. ex

Wild & R. B. Drumm.; Verdcourt, F.T.E.A. (Vitaceae): 81. 1993.

Terrestrial herb. Miombo woodland. 1200–1530 m.

Distribution: MP, SL—*Gereau 3834*

Cyphostemma cyphopetalum (Fresen.) Desc. ex Wild & R. B. Drumm.; Verdcourt, F.T.E.A. (Vitaceae): 121. 1993.

Herbaceous climber. Montane forest, woodland, abandoned cultivation. 1250–2350 m.

Distribution: MP, MR, RM—*Gereau 4601*

Cyphostemma gigantophyllum (Gilg & M. Brandt) Desc. ex Wild & R. B. Drumm.; Verdcourt, F.T.E.A. (Vitaceae): 68. 1993.

Herbaceous climber. Riverine forest, miombo woodland. 1500–1620 m.

Distribution: SL—*Gereau 3939*

Cyphostemma heterotrichum (Gilg & R. E. Fr.) Desc. ex Wild & R. B. Drumm.; Verdcourt, F.T.E.A. (Vitaceae): 77. 1993.

Terrestrial herb. Woodland, abandoned cultivation. 1200–1500 m.

Distribution: MP—*J. Lovett 4356*

Cyphostemma hildebrandtii (Gilg) Desc. ex Wild & R. B. Drumm.; Verdcourt, F.T.E.A. (Vitaceae): 66. 1993.

Herbaceous climber. Riverine forest. 580 m.

Distribution: SL—*Gereau 3816*

Cyphostemma hypoleucum (Harv.) Desc. ex Wild & R. B. Drumm. vel sp. aff.; Wild & Drummond, Fl. Zamb. 2: 484. 1966.

Herbaceous climber. Scree slopes. 2060–2120 m.

Distribution: SL—*Gereau 4439*

Cyphostemma junceum (Webb) Desc. ex Wild & R. B. Drumm. subsp. *jatrophoides* (Welw. ex Baker) Verdc.; Verdcourt, F.T.E.A. (Vitaceae): 62. 1993.

Terrestrial herb. Woodland, edges of cultivation. 1300 m.

Distribution: MP—*J. Lovett 3552*

Cyphostemma kilimandscharicum (Gilg) Desc. ex Wild & R. B. Drumm. var. *jaegeri* (Gilg & M. Brandt) Verdc.; Verdcourt, F.T.E.A. (Vitaceae): 110. 1993.

Herbaceous climber. Montane forest. 1790 m.

Distribution: RM—*Abeid 2917*

Cyphostemma kilimandscharicum (Gilg) Desc. ex Wild & R. B. Drumm. var. *kilimandscharicum*; Verdcourt, F.T.E.A. (Vitaceae): 108. 1993.

Herbaceous climber or liana. Montane forest, tree clumps in grassland. 1730–2400 m.

Distribution: MP, MR, NL—*Gereau 4144*

Cyphostemma masukuense (Baker) Desc. ex Wild & R. B. Drumm. subsp. *masukuense*; Verdcourt, F.T.E.A. (Vitaceae): 105. 1993.

Herbaceous climber or weak liana. Montane forest, woodland. 1950–2110 m.

Distribution; MP, MR, RM—*Gereau 4554*

Cyphostemma princeae (Gilg & M. Brandt) Desc. ex Wild & R. B. Drumm.; Verdcourt, F.T.E.A. (Vitaceae): 79. 1993.

Terrestrial herb. Woodland, riverine forest, abandoned cultivation. 1200–1720 m.

Distribution: MP—*J. Lovett 4078*

Cyphostemma pseudosesquipedale Verdc.; Verdcourt, F.T.E.A. (Vitaceae): 83. 1993.

Terrestrial herb. Woodland. 1200 m.

Distribution: MP—*J. Lovett 3953*

Cyphostemma vandenbrandeanum (Dewit) Desc. ex Wild & R. B. Drumm.; Verdcourt, F.T.E.A. (Vitaceae): 102. 1993.

Herbaceous climber. Scree slopes, rocky mountain grassland. 1980–2140 m.

Distribution: SL—*Gereau 3643*

Cyphostemma sp.

Herbaceous climber. Riverine forest. 1210 m.

Distribution: SL—*Gereau 4254*

Rhoicissus tridentata (L. f.) Wild & R. B. Drumm.; Verdcourt, F.T.E.A. (Vitaceae): 9. 1993.

Liana. Montane grassland, montane forest edges, thickets, riverine forest, woodland, abandoned cultivation. 1100–2140 m.

Distribution: MP, NL, RM, SL—*Suleiman 96*

XYRIDACEAE

1 genus and 1 species

Xyris obscura N. E. Br.; Lock, F.T.E.A. (Xyridaceae): 10. 1999.

Terrestrial herb. Wet montane grassland. 2550–2740 m.

Distribution: KP, RM—*Gereau 3187*

ZINGIBERACEAE

2 genera and 2 species

Aframomum zambesiacum (Baker) K. Schum.; Lock, F.T.E.A. (Zingiberaceae): 25. 1985.

Terrestrial herb. Stream banks in woodland. 1650–1680 m.

Distribution: RM, SL—*Gereau 5120*

Siphonochilus rhodesicus (T. C. E. Fr.) Lock; Lock, F.T.E.A. (Zingiberaceae): 20. 1985.

Terrestrial herb. Miombo woodland. 1200 m.

Distribution: MP—*J. Lovett 4677*

ZYGOPHYLLACEAE

1 genus and 1 species

Balanites aegyptiacus (L.) Delile; Sands, F.T.E.A. (Balanitaceae): 6. 2003.

Tree. Woodland, termite mounds in grassland. 1250–1400 m.

Distribution: MP—*J. Lovett 4916*

Literature Cited

Asfaw Hunde & M. Thulin. 1989. Fabaceae–Mimosoideae. Pp. 71–96 *in* I. Hedberg & S. Edwards (editors), Flora of Ethiopia, Vol. 3. The National Herbarium, Addis Ababa University, Addis Ababa, and The Department of Systematic Botany, Uppsala University, Uppsala.

Atlas of Tanzania. 1967. Government Surveys and Mapping Division, Ministry of Lands, Settlements and Water Development. Government Printer, Dar es Salaam.

Bamps, P. 1974. Contribution à l'étude des Araliacées africaines. Bull. Jard. Bot. Belg. 44: 101–139.

Barnes, R. D. & C. W. Fagg. 2003. *Faidherbia albida*: Monograph and annotated bibliography. Oxford Forestry Inst. Trop. Forestry Papers 41: i–xiv, 1–267.

Beentje, H. J. (editor). 1998–2007. Flora of Tropical East Africa. A. A. Balkema, Rotterdam.

Beentje, H. J. (editor). 2010. Flora of Tropical East Africa. Royal Botanic Gardens, Kew.

Beentje, H. J. & S. Smith. 2001. FTEA and after. Pp. 265–290 *in* E. Robbrecht, J. Degreef & I. Friis (editors), Proceedings of the XVIth AETFAT Congress. Syst. Geogr. Pl. 71.

Beentje, H. J. & S. A. Ghazanfar (editors). 2005–2009. Flora of Tropical East Africa. Royal Botanic Gardens, Kew.

Brenan, J. P. M. (editor). 1970. Flora Zambesiaca, Vol. 3(1). Crown Agents for Oversea Governments and Administrations, London.

Bruce, E. A. 1937. *Raphionacme longituba* E. A. Bruce. Pp. 419–420 *in* M. B. Moss, E. Milne-Redhead, E. A. Bruce, R. A. Dyer, J. Hutchinson, P. J. Greenway, M. R. F. Taylor, A. A. Bullock, H. L. Dunkley & B. L. Burtt, Tropical African Plants: XVIII. Bull Misc. Inform. 1937.

Brummitt, R. K. 2000. Inclusion of *Clematopsis* Hutch. in *Clematis* L. (Ranunculaceae). Kew Bull. 55: 97–108.

Bullock, A. A. 1952. Notes on African Asclepiadaceae. I. Kew Bull. 1952: 405–426.

Bullock, A. A. 1953a. Notes on African Asclepiadaceae. II. Kew Bull. 1953: 51–67.

Bullock, A. A. 1953b. Notes on African Asclepiadaceae. III. Kew Bull. 1953: 329–362.

Burgess, N., J. Lovett, W. A. Rodgers, F. Kilahama, T. R. B. Davenport & T. Butynski. 2004. Eastern Arc Mountains and Southern Rift. Pp. 245–255 *in* R. A. Mittermeier, R. Robles G., M. Hoffman, J. Pilgrim, T. Brooks, C. G. Mittermeier, J. Lamoureux & G. A. B. DaFonseca, Hotspots Revisited. CEMEX, Mexico City.

Clayton, W. D. 1985. Miscellaneous notes on Pooid grasses. Kew Bull. 40: 727–729.

Cribb, P. J. & G. P. Leedal. 1982. The Mountain Flowers of Southern Tanzania. A. A. Balkema, Rotterdam.

Cribb, P. J. & H. Kurzweil. 1994. The genus *Corycium* (Orchidaceae) newly reported for Tanzania. Kew Bull. 49: 555–557.

Cronquist, A. 1981. An Integrated System of Classification of Flowering Plants. Columbia University Press, New York.

Dahlgren, R. 1985. The Families of the Monocotyledons: Structure, Evolution, and Taxonomy. Springer-Verlag, Berlin.

Davenport, T. R. B. 2002. Tanzania's new national park to protect orchids. Oryx 36: 224.

Davenport, T. R. B. 2004. Where the arcs meet the rift: Biogeography, outliers and conservation in Tanzania's Southern Highlands. Soc. Conserv. Biol. Abstr. 48(5): 20.

Davenport, T. R. B. 2005. Merely "Other forested mountains"? Biodiversity, biogeography and conservation in Tanzania's Southern Highlands. Proceedings of the Fifth Annual Scientific Conference, Tanzania Wildlife Research Institute, Arusha (unpublished).

Davenport, T. R. B. 2006. Plants, primates and people: Conservation in the Southern Highlands of Tanzania. Miombo 28: 7–8.

Davenport, T. R. B. & H. J. Ndangalasi. 2003. An escalating trade in orchid tubers across Tanzania's Southern Highlands: Assessment, dynamics and conservation implications. Oryx 37: 55–61.

Davenport, T. R. B. & B. Bytebier. 2004. Kitulo Plateau: A first park for orchids. Orchid Rev. 112: 160–165.

Davenport, T. R. B., W. T. Stanley, E. J. Sargis, D. W. De Luca, N. E. Mpunga, S. J. Machaga & L. E. Olson. 2006. A new genus of African monkey, *Rungwecebus*:

Morphology, ecology and molecular phylogenetics. Science 312: 1378–1381.

Davenport, T. R. B., D. W. De Luca, T. Jones, N. E. Mpunga, S. J. Machaga, A. Kitegile & G. P. Picton Phillipps. 2008. The critically endangered kipunji *Rungwecebus kipunji* of southern Tanzania: First census and conservation status assessment. Oryx 42: 352–359.

Dietrich, W., W. L. Wagner & P. H. Raven. 1997. Systematics of *Oenothera* section *Oenothera* subsection *Oenothera* (Onagraceae). Syst. Bot. Monogr. 50: 1–234.

Dyer, W. T. T. (editor). 1896–1913. Flora of Tropical Africa, Vol. 4–6(1), 7–8. L. Reeve & Co., London.

Elton, J. F. & H. B. Cotteril. 1879. Travels and Researches among the Lakes and Mountains of Eastern and Central Africa. John Murray, London.

Engler, A. & W. Goetze. 1902. Vegetationsansichten aus Deutschostafrika, insbesondere aus der Khutusteppe, dem Ulugurugebirge, Uhehe, dem Kingagebirge, vom Rungwe, dem Kondeland und der Rukwasteppe. W. Engelmann, Leipzig.

Ensermu Kelbessa & R. B. Faden. 1997. Commelinaceae. Pp. 339–377 *in* S. Edwards, Sebsebe Demissew & I. Hedberg (editors), Flora of Ethiopia and Eritrea, Vol. 6. The National Herbarium, Addis Ababa University, Addis Ababa, and The Department of Systematic Botany, Uppsala University, Uppsala.

Exell, A. W. & H. Wild (editors). 1960–1961. Flora Zambesiaca, Vol. 1(1, 2). Crown Agents for Oversea Governments and Administrations, London.

Exell, A. W. & E. Launert (editors). 1970. Flora Zambesiaca, Pteridophyta. Crown Agents for Oversea Governments and Administrations, London.

Exell, A. W., A. Fernandes & H. Wild (editors). 1963, 1966. Flora Zambesiaca, Vol. 2(1, 2). Crown Agents for Oversea Governments and Administrations, London.

Faden, R. B. 1991. The morphology and taxonomy of *Aneilema* R. Brown (Commelinaceae). Smithsonian Contr. Bot. 76: i–vi, 1–166.

Faden, R. B. 1994. Commelinaceae. Pp. 303–309 *in* A. D. Q. Agnew & S. Agnew, Upland Kenya Wild Flowers, 2nd ed. East Africa Natural History Society, Nairobi.

Faden, R. B. 2001. New or misunderstood species of *Commelina* (Commelinaceae) from the *Flora of Tropical East Africa* and *Flora Zambesiaca* areas. Novon 11: 398–409.

Fernandes, A., E. Launert & H. Wild (editors). 1971. Flora Zambesiaca, Vol. 10(1). Crown Agents for Oversea Governments and Administrations, London.

Fernandes, R. B. & B. Verdcourt. 2000. *Rotheca* (Labiatae) revived—More new combinations. Kew Bull. 55: 147–154.

Frimodt-Møller, C. & C. Grey-Wilson. 1999. Two new taxa of *Impatiens* (Balsaminaceae) from the Udzungwa Mountains, Tanzania. Kew Bull. 54: 179–184.

Geerinck, D. 1984. Orchidaceae (première partie). *In* P. Bamps (editor), Flore d'Afrique Centrale (Zaïre–Rwanda–Burundi). Jardin botanique national de Belgique, Brussels.

Gillman, C. 1927. South-west Tanganyika Territory. Geogr. J. (London) 69: 9–131.

Govaerts, R. 1996. World Checklist of Seed Plants, Vol. 2. Continental Publishing, Antwerp.

Goyder, D. J. 1990. A new species of *Brachystelma* (Asclepiadaceae) from southern Tanzania. Kew Bull. 45: 729–731.

Goyder, D. J. 2003. *Xysmalobium* R. Br. Pp. 131–133 *in* I. Hedberg, S. Edwards & Sileshi Nemomissa (editors), Flora of Ethiopia and Eritrea, Vol. 4, part 1. The National Herbarium, Addis Ababa University, Addis Ababa, and The Department of Systematic Botany, Uppsala University, Uppsala.

Greenway, P. J. 1955. Ecological observations on an extinct East African volcanic mountain. J. Ecol. 43: 544–563.

Haines, R. W. & K. A. Lye. 1983. The Sedges and Rushes of East Africa. East Africa Natural History Society, Nairobi.

Hall, A. V. 1999. *Satyrium* Sw. Pp. 152–181, pl. 137–180 *in* H. P. Linder & H. Kurzweil, Orchids of Southern Africa. A. A. Balkema, Rotterdam.

Harkin, D. A. 1960. The Rungwe Volcanics at the Northern End of Lake Nyasa. Geological Survey of Tanganyika Memoir 2. Government Printer, Dar es Salaam.

Harpum, J. R. 1970. Summary of the Geology of Tanzania. Part V: Structure and Geotechnics of the Precambrian. Memoir 1. Mineral Resources Division, Ministry of Commerce and Industry. Government Printer, Dodoma, Tanzania.

Hedberg, O. 1957. Afroalpine vascular plants. A taxonomic revision. Symb. Bot. Upsal. 15: 1–411, pl. 1–12.

Hepper, F. N. (editor). 1963–1972. Flora of West Tropical Africa, 2nd ed., Vol. 2, 3(1, 2). Crown Agents for Oversea Governments and Administrations, London.

Hiepko, P. 1989. Opiliaceae. Pp. 358–359 *in* I. Hedberg & S. Edwards, Flora of Ethiopia, Vol. 3. The National Herbarium, Addis Ababa University, Addis Ababa, and The Department of Systematic Botany, Uppsala University, Uppsala.

Hill, A. W. (editor). 1937. Flora of Tropical Africa, Vol. 10(1). L. Reeve & Co., London.

Hilu, K. W. 1994. Validation of the combination *Eleusine coracana* subspecies *africana* (Kennedy-O'Byrne) Hilu & de Wet. Phytologia 76: 410–411.

Hong, S. P. 1988. A pollen morphological re-evaluation of *Harpagocarpus* and *Eskemukerjea* (Polygonaceae). Grana 27: 291–295.

Hooper, S. S. 1959. The genus *Dianthus* in Central and South Africa. Hooker's Icon. Pl. 37: 1–60, tab. 3601–3624.

Hubbard, C. E. & E. Milne-Redhead (editors). 1958–1966. Flora of Tropical East Africa. Crown Agents for Oversea Governments and Administrations, London.

Huber, H. 1957. Revision der Gattung *Ceropegia*. Mem. Soc. Brot. 12: 5–203, tab. 1–16.

Hutchinson, J. & J. M. Dalziel. 1954, 1958. Flora of West Tropical Africa (revised by R. W. J. Keay), 2nd ed., Vol. 1 (1, 2). Crown Agents for Oversea Governments and Administrations, London.

Ittenbach, S. 2003. Revision der afrikanischen Arten der Gattung *Amorphophallus* (Araceae). Englera 25: 1–263.

Jones, T., C. Ehardt, T. Butynski, T. R. B. Davenport, N. E. Mpunga, S. J. Machaga & D. W. De Luca. 2005. *Lophocebus kipunji*: A new species of African monkey. Science 308: 1161–1164.

Jongkind, C. C. H. 1989. *Rourea* Aublet. Pp. 310–368 *in* F. J. Breteler (editor), The Connaraceae: A Taxonomic Study with Emphasis on Africa. Agric. Univ. Wageningen Pap. 89(6).

Jordaan, M. & A. E. van Wyk. 2004 [2003]. A taxonomic revision of the *Gymnosporia mossambicensis* group (Celastraceae: Celastroideae). Kew Bull. 58: 833–866.

Jordaan, M. & A. E. van Wyk. 2006. Sectional classification of *Gymnosporia* (Celastraceae), with notes on the nomenclatural and taxonomic history of the genus. Taxon 55: 515–525.

Kamukala, G. L. & S. A. Crafter (editors). 1993. Wetlands of Tanzania. Proceedings of a Seminar on the Wetlands of Tanzania, Morogoro, Tanzania, 27–29 November 1991.

Kårehed, J. & B. Bremer. 2007. The systematics of Knoxieae (Rubiaceae)—Molecular data and their taxonomic consequences. Taxon 56: 1051–1076.

Kerfoot, O. 1964. A preliminary account of the vegetation of the Mbeya Range, Tanganyika. Kirkia 4: 191–206.

Kubitzki, K. 1990. The Families and Genera of Vascular Plants, Vol. 1. Pteridophytes and Gymnosperms (volume editors K. U. Kramer & P. S. Green). Springer-Verlag, Berlin.

Lantz, H. & B. Bremer. 2004. Phylogeny inferred from morphology and DNA data: Characterizing well-supported groups in Vanguerieae (Rubiaceae). Bot. J. Linn. Soc. 146: 257–283.

Launert, E. (editor). 1978–1989. Flora Zambesiaca, Vol. 4, 7(1, 2), 8(1, 3), 9(1), 5(1). Flora Zambesiaca Managing Committee, London.

Launert, E. & G. V. Pope (editors). 1989–1991. Flora Zambesiaca, Vol. 10(3), 7(4), 8(2), 9(6). Flora Zambesiaca Managing Committee, London.

Leeuwenberg, A. J. M. 1979. The Loganiaceae of Africa XVIII. *Buddleja* L. II. Revision of the African and Asiatic species. Meded. Landbouwhogeschool 79(6): 1–163.

Liede, S. 1996a. A revision of *Cynanchum* (Asclepiadaceae) in Africa. Ann. Missouri Bot. Gard. 83: 283–345.

Liede, S. 1996b. *Cynanchum–Rhodostegiella–Vincetoxicum–Tylophora* (Asclepiadaceae): New considerations on an old problem. Taxon 45: 193–211.

Linder, H. P. 1981. Taxonomic studies on the Disinae. III. A revision of *Disa* Berg. excluding sect. *Micranthae* Lindl. Contr. Bolus Herb. 9: 1–370.

Linder, H. P. 1999. *Disa* Berg. Pp. 187–288, pl. 186–314 *in* H. P. Linder & H. Kurzweil, Orchids of Southern Africa. A. A. Balkema, Rotterdam.

Lisowski, S. 1989. Compositae (Deuxième partie: tribu Inuleae). *In* P. Bamps (editor), Flore d'Afrique Centrale, Spermatophytes. Jardin botanique national de Belgique, Brussels.

Lourteig, A. 1981. *Biophytum umbraculum* (Oxalidaceae). Brittonia 33: 451–452.

Lovett, J. C. & E. Prins. 1994. Estimation of land use changes on the Kitulo Plateau using satellite imagery. Oryx 28: 173–182.

Lovett, J. C., R. E. Gereau & K. J. Sidwell. 1994. Vegetation and phytogeography of the Kitulo Plateau, southern Tanzania. Pp. 1025–1041 *in* J. H. Seyani & A. C. Chikuni, Proceedings of the XIIIth Plenary Meeting of AETFAT, Zomba, Malawi, 2–11 April 1991. National Herbarium and Botanic Gardens of Malawi, Zomba.

Lovett, J. C., C. K. Ruffo, R. E. Gereau & J. R. D. Taplin. 2006. Field Guide to the Moist Forest Trees of Tanzania. The Society for Environmental Exploration, London, and the University of Dar es Salaam, Dar es Salaam.

Machaga, S. J., A. A. Massawe & T. R. B. Davenport. 2004. Medicine, meat & mahogany: Understanding natural resource use on Mt. Rungwe, SW Tanzania. Soc. Conserv. Biol. Abstr.

Machaga, S. J., A. A. Massawe & T. R. B. Davenport. 2005. Food, fuel and fitness. Quantifying natural resource use on Mt. Rungwe. TAWIRI Abstr.

Manning, J. C. & P. Goldblatt. 2009. Three new species of *Gladiolus* (Iridaceae) from South Africa, a major range extension for *G. rubellus* and taxonomic notes for the genus in southern and tropical Africa. Bothalia 39: 37–45.

Milne-Redhead, E. & R. M. Polhill (editors). 1966–1972. Flora of Tropical East Africa. Crown Agents for Oversea Governments and Administrations, London.

Moffett, J. P. (editor). 1958. Handbook of Tanganyika, 2nd ed. Government Printer, Dar es Salaam.

Moffett, R. O. 1993. Anacardiaceae: *Rhus*. Pp. iii–v, viii, 1–129 *in* O. A. Leistner (editor), Flora of Southern Africa, Vol. 19(3/1). Botanical Research Institute, Pretoria.

Moore, S. Le M. 1908. Alabastra Diversa.—Part XVII. J. Bot. 46: 290–298, 305–313.

Mwakilema, W. S. & T. R. B. Davenport. 2005. Managing Kitulo—How to protect Tanzania's newest and most unique national park. TAWIRI Abstr.

Napper, D. M. 1965. Grasses of Tanzania, with Keys to Identification. Ministry of Agriculture, Forests and Wildlife. Government Printer, Dar es Salaam.

NORPLAN. 2002. Lower Kihansi Hydropower Project. Immediate Rescue and Emergency Measures. Final Main Report: Components 1–7. Report produced for Tanzania Electric Supply Company Limited (TANESCO), Dar es Salaam.

Ntore, S., E. Robbrecht, E. Smets & S. Dessein. 2003. Deux nouvelles espèces paléo-endémiques de *Pauridiantha* (Rubiaceae) des Monts Udzungwa (sud de la Tanzanie). Adansonia, sér. 3, 25: 65–76.

Oliver, D. (editor). 1868–1877. Flora of Tropical Africa, Vol. 1–3. L. Reeve & Co., London.

Pasquet, R. S. 2007. *Vigna verdcourtii* (Papilionoideae), a new species from eastern Africa. Bothalia 37: 51–54.

Phillips, S. 1995. Poaceae (Gramineae). Pp. 1–397 *in* I. Hedberg & S. Edwards (editors), Flora of Ethiopia and Eritrea, Vol. 7, The National Herbarium, Addis Ababa University, Addis Ababa, and The Department of Systematic Botany, Uppsala University, Uppsala.

Platts, P. J., N. D. Burgess, R. E. Gereau, J. C. Lovett, A. R. Marshall, C. J. McLean, P. K. E. Pellikka, R. D. Swetnam & R. Marchant. 2011. Delimiting tropical mountain ecoregions for conservation. Environm. Conservation 38: 312–324.

Polhill, R. M. (editor). 1973–1978. Flora of Tropical East Africa. Crown Agents for Oversea Governments and Administrations, London.

Polhill, R. M. 1982. *Crotalaria* in Africa and Madagascar. A. A. Balkema, Rotterdam.

Polhill, R. M. (editor). 1982–1997. Flora of Tropical East Africa. A. A. Balkema, Rotterdam.

Pope, G. V. (editor). 1992–1995. Flora Zambesiaca, Vol. 6(1), 12(4), 11(1). Flora Zambesiaca Managing Committee, London.

Pope, G. V. (editor). 1996–2005. Flora Zambesiaca, Vol. 9(4), 5(2, 3), 11(2), 10(2), 3(5, 6, 7), 8(4, 7), 9(5), 10(4), 12(3). Royal Botanic Gardens, Kew.

Pope, G. V. & E. S. Martins (editors). 2005. Flora Zambesiaca, Vol. 8(4). Royal Botanic Gardens, Kew.

Pope, G. V., R. M. Polhill & E. S. Martins (editors). 2006. Flora Zambesiaca, Vol. 9(3). Royal Botanic Gardens, Kew.

Power, S., P. Makungu & A. Qaraeen (editors). 1997. Watershed Management, Participatory and Sustainable Development. Proceedings of DANIDA's Second International Workshop on Watershed Development, Iringa Region, Tanzania, 26 May to 5 June 1997.

Prain, D. (editor). 1916–1934. Flora of Tropical Africa, Vol. 6(2), 9. L. Reeve & Co., London.

Prain, D. & J. Hutchinson. 1913. Notes on some species of *Acalypha*. Bull. Misc. Inform. 1913: 1–28.

Prance, G. T. 2002. New combinations in African *Chrysobalanaceae*. Kew Bull. 57: 993–995.

Prance, G. T. & F. White. 1988. The genera of Chrysobalanaceae: A study in practical and theoretical taxonomy and its relevance to evolutionary biology. Philos. Trans., Ser. B 320: 1–184.

Pratt, D. J. & M. D. Gwynne (editors). 1977. Rangeland Management and Ecology in East Africa. Robert E. Krieger, Huntington, New York.

Prins, W. H. & J. Prins-Lampert. 1974. A first check list of vascular plants of the Kitulo Plateau, southern highlands of Tanzania. Tanzania Notes Rec. 73: 35–42.

Ridsdale, C. E. 1976. A revision of the tribe *Cephalantheae* (Rubiaceae). Blumea 23: 177–188.

Robbrecht, E. 1979. The African genus *Tricalysia* A. Rich. (Rubiaceae–Coffeeae). 1. A revision of the species of subgenus *Empogona*. Bull. Jard. Bot. Belg. 49: 239–360.

Robbrecht, E. 1982. The African genus *Tricalysia* A. Rich. (Rubiaceae–Coffeeae). 2. *Ephedranthera*, a new section of subgenus *Tricalysia*. Bull. Jard. Bot. Belg. 52: 311–339.

Schrire, B. D. 1992. New combinations and resurrected names in *Microcharis* and *Indigastrum* (Fabaceae–Papilionoideae). Bothalia 22: 165–170.

Sleumer, H. 1972. A taxonomic revision of the genus *Dovyalis* E. Mey. ex Arn. (Flacourtiaceae). Bot. Jahrb. Syst. 92: 64–89.

Smith, D. M. N. 1988. A revision of the genus *Pachycarpus* in southern Africa. S. African J. Bot. 54: 399–439.

Sonder, W. 1860. Caryophylleae. Pp. 120–151 *in* W. H. Harvey & O. W. Sonder, Flora Capensis, Vol. 1. Hodges, Smith, and Co., Dublin.

Staddon, S., C. Bracebridge, E. Fanning & K. M. Howell. 2003. Mpanga/Kipengere Game Reserve: A Biodiversity Study. Frontier Tanzania Environmental Research Report 99. Society for Environmental Exploration UK, University of Dar es Salaam, Wildlife Division (WD), World Wide Fund for Nature (WWF), Dar es Salaam.

Stafleu, F. A. & R. S. Cowan. 1981. Taxonomic Literature, 2nd ed., Vol. 3, Lh–O. Regnum Veg. 105: v–12, 1–980.

Timberlake, J. R. & E. S. Martins (editors). 2008–2010. Flora Zambesiaca, Vol. 13(1), 12(2), 13(2, 4). Royal Botanic Gardens, Kew.

Timberlake, J. R., G. V. Pope, R. M. Polhill & E. S. Martins (editors). 2007. Flora Zambesiaca, Vol. 3(2, 3). Royal Botanic Gardens, Kew.

TISCO. 1983. Dairy Development for Kitulo Plateau. ISIC No. 3112, TISCO Project 82–050.

Turner, B. L. & M. G. Mendenhall. 1993. A revision of *Malvaviscus* (Malvaceae). Ann. Missouri Bot. Gard. 80: 439–457.

Turrill, W. B. & C. E. Milne-Redhead (editors). 1952–1958. Flora of Tropical East Africa. Crown Agents for Oversea Governments and Administrations, London.

Venter, H. J. T. 2003. *Periploca* L., *Tacazzea* Decne. Pp. 657–660 *in* I. Hedberg, S. Edwards & Sileshi Nemomissa, Flora of Ethiopia and Eritrea, Vol. 4(1). The National Herbarium, Addis Ababa University, Addis Ababa, and The Department of Systematic Botany, Uppsala University, Uppsala.

Verdcourt, B. 1987. Three corrections to the Flora of Tropical East Africa. Kew Bull. 42: 657–660.

Weimarck, H. 1940. Monograph of the genus *Aristea*. Acta Univ. Lund, n.s. 36: 1–140.

White, F. 1983. The Vegetation of Africa. A Descriptive Memoir to Accompany the Unesco/AETFAT/UNSO Vegetation Map of Africa. Unesco, Paris.

Wilczek, R. 1952. Bauhinieae. Pp. 265–278 *in* Comité exécutive de la Flore du Congo belge, Flore du Congo belge et du Ruanda—Urundi, Vol. 3. I.N.É.A.C., Brussels.

Winter, P. J. D. & B.-E. van Wyk. 1996. A revision of the genus *Heteromorpha* (Apiaceae). Kew Bull. 51: 225–265.

Zizka, G. 1988. Revision der Melinideae Hitchcock (Poaceae, Panicoideae). Biblioth. Bot. 138: 1–149.